*Special Relativity
is
Nonsense*

*It is vain to do with more what can be done with less*

William of Ockham
1285-1349

*Simplex sigillum veri*

Herman Boerhaave
1668-1738

*ex nihilo nihilum*

Parmenides
500 BC

# Jan Slowak

# *Special Relativity is Nonsense*

## Previous books

1. Bye-Bye Big Bang,
   Episod/Episode 1, 2, 3; swedish-english
4. Redshift factor, Absolute redshift,
   Galaxies red / blue distribution
5. Sawing of my article about the Big Bang
6. Big Bang -
   Questions to physicists and cosmologists
7. Einstein's special theory of relativity -
   mathematical and physical mistakes!
9. Back to Newton, swedish, english
10. Einstein's special theory of relativity =
    mathematical and ... swedish
11. SR(LT, LF, TD, LK) = NONSENS, swedish

Copyright © Jan Slowak 2020
Publisher: Books on Demand, Stockholm, Sweden
Printed: Books on Demand, Norderstedt, Germany
ISBN: 978-91-7785-965-9

*Jan Slowak: Special Relativity is Nonsense*

*For
Science*

## Content

Bibliography ... 7
What is the special theory of relativity? ... 9
Prologue ... 11   Historical review ... 15
Everything is relative ... 16
Events in coordinate system ... 17   Light ... 18
Registration, calculation and transformation of coordinates ... 20
Analysis of time dilation ... 26
Derivation of Lorentz transformations 1 - 6 ... 34
Michelson-Morley experiment ... 51
Mathematics and SR ... 71
Physics and SR ... 73
Special theory of relativity and LT ... 77
Relativity with classical physics ... 83
Derivation of LT ... 90   Verification of LT ... 97
SR and spacetime ... 107
Lorentz factor and its value ... 113
The twin paradox ... 118
The twin paradox: The third brother ... 135
My patent application: 1ANM ... 145
Ending ... 171
Quotes from books I read ... 174
To physicists and mathematicians ... 198
Articles submitted for publication... 200
Publiched articles... 207
Some of my articles ... 211

# Bibliography

[1] Modern Physics; Sixth edition; Paul A. Tipler, Ralph A. Llewellyn; Chapter 1; Relativity I; 2012

[2] University Physics with Modern physics; Thirteen Edition; Young Freedman; Chapter 37; Relativity; 2012

[3] Den speciella och den allmänna relativitetsteorin; Albert Einstein; Första delen; Om den speciella relativitetsteorin; 2006; swedish

[4] Einsteins relativitetsteori – en kritisk analys ...; Ove Tedenstig; 2015; swedish

[5] Den moderna fysikens grunder ...; Krister Renard; Kapitel 2; Speciell relativitetsteori; 1995; swedish

[6] Concepts of Modern Physics; Sixth edition; Arthur Beiser; Chapter 1; Relativity; 2003

[7] Modern Physics; Second edition; Randy Harris; Chapter 2; Special Relativity; 2008

[8] Knowing, The Nature of Physical law, Michael Munowitz, 2005

[9] Illustrerad vetenskap, Nr 16/2014; swedish

[10] Calculus - A Complete Course; Robert A. Adams; Sixth Edition;

[11] Nádherná teorie – Sto let obecné teorie relativity; Pedro G. Ferreira; czech

[12] Six Ideas That Shaped Physics; Thomas A. Moore; 2003

[13] Calculating the cosmos; Ian Stewart; 2016

[14] Rumtid – en introduktion till Einsteins relativitetsteori; Sören Holst; 2006; swedish

[15] Relativitet – Teorin som revolutionerade vår syn på universum; Jeffrey Bennett; 2015; swedish

[16] Mörk energi = Gravitation; Robin T. Trnovsky; 2016; swedish

[17] Det europeiska miraklet; Bok 1; Robin T. Trnovsky; 2016; swedish

[18] Einsteins största misstag - Ett geni med fel och brister; David Bodanis; 2017; swedish
... and more ...

## What is the special theory of relativity?

The special theory of relativity is a physical theory published in 1905 by Albert Einstein.
This theory describes the properties of space and time (?) and the relationship of events in so-called **inertial systems**.

According to the special theory of relativity, the space ($x$, $y$, $z$) and time ($t$) together form a four-dimensional system, so-called **spacetime** ($x$, $y$, $z$, $t$), where **distance and time measurements depend on the observer's motion**.

According to this theory, there are no **absolute motions** (?) or time course, but these are relative and the speed of an object can only be set relative to other objects (?).

The special theory of relativity also states that there is a maximum speed, **the speed of light in vacuum**, and that this speed is **constant and equal for all observers**.

Objects moving relative to the observer are shortened in the direction of motion (?), according to the observer's measurements in this direction. No local

contraction of the object occurs (!).
**Clocks in motion go slower than clocks in rest (?).**

The special theory of relativity describes two inertial systems that are at rest (?) or move at a constant speed relative to each other. An event in such a system can be denoted by $E = (x, y, z, t)$. The transition from an inertial system, S', to another, S, and vice versa, is done using Lorentz transformations.

$$x' = (x - vt)\gamma \qquad (LT_1)$$
$$t' = (t - vx/c^2)\gamma \qquad (LT_2)$$

where $\gamma = 1/(1 - v^2/c^2)^{1/2}$ is called the Lorentz factor.

You can see that I have marked some statements with question marks. These are claims that relate to the special theory of relativity that I cannot accept. This is what I have tried to disprove.

Follow my work in which I come to the conclusion that I state in the book's title!

# Prologue

When I started researching for real the theory of special relativity, I went thoroughly through most of the books I found in the university library. I mean that I read everything that concerned the theory of special relativity. Then it was the articles on the web. Some of those told that all scientists do not agree on this theory.

It is not easy to research a subject that has nothing in common with your real job! In any case, all my spare time went for this purpose. One may wonder why I was doing so. For me it was an old desire. The first time I got in touch with the theory of relativity was in high school. And I could *not* accept it. Not time dilation, not the twins paradox, not the length contraction.

The time went by. I studied mathematics and computer science at university. And since then, I have worked as a software developer, a programmer.

So why should *I* do research on the theory of special relativity? Why so late? Well, I was always overwhelmed by the science. I was overwhelmed every time I was reading about scientists who came up with

new findings and explanations how things work in different fields: anthropology, genetics, astronomy, cosmology.
No, not cosmology, not the expanding of the universe, not the Big Bang, not the dark matter.

But why could I accept most of the new ideas except those in cosmology?
Because my motto was what we learned in school:

*ex nihilo nihil fit*

I started my research in cosmology and the theory of special relativity sometime in 2014. It was the analysis of data from the database NED Redshift-Independent Distances from
http://ned.ipac.caltech.edu/Library/Distances/

The result of this research I published in my book *Redshift factor, Absolute redshift, Galaxies red / blue distribution*. And the result was astounding, in my opinion:

| Population | zf | Num obj | Num red z | % red z | Num blue z | % blue z |
|---|---|---|---|---|---|---|
| NED-D | 0.000239 | 26,790 | 13,018 | 48.6 | 13,772 | 51.4 |

We see here that the distribution of the objects' redshift and blue shift is about 50/50!
The Big Bang theory says that *most* cosmic object has a redshift, except for some in our neighborhood that may have blue shift. My research shows that there is no argument for the expansion of the universe!

I sent my book to some researchers. The book was sawed! One could say that the result was based on my own interpretation of the data.

Therefore, I decided to go to the source of the problem. Big Bang theory was based on Einstein's theory of relativity.

The result of this research, I published in my book *Einstein's theory of special relativity - mathematical and physical mistakes!* (swedish)

But today, anyone can publish a book. The question is whether you get recognition for your ideas and your research. It is a difficult task! It's like fighting against the "windmills"!
I wrote a few articles based on my book and sent them to some magazines. I sent question to a number of institutions if I could present my research to some researchers. I got no answer or the answer was negative!

In this book I intend to summarize my research on the theory of special relativity. I will come up with evidence that the theory of special relativity is wrong fundamentally, in its entirety!

# Historical review

We present some scientists that somehow were mentioned when talking about the theory of special relativity.

### Galileo Galilei, 1564-1642
Galilean transformation, $x' = x - vt$

### Isaac Newton, 1642-1727
Time is universal and the same everywhere; Space is the same in all places, and the same in all directions; Space is homogeneous and isotropic; Absolute motion

### James Clerk Maxwell, 1831-1879
Maxwell equations; The speed of light, $c = 1/(\mu_0 \varepsilon_0)^{1/2}$; The light-bearing ether

### Albert A. Michelson, 1852-1931
Michelson-Morley experiment, 1887

### Hendrik A. Lorentz, 1853-1928
Lorentz transformation, $x' = (x - vt)\gamma$, $t' = (t - vx/c^2)\gamma$; Lorentz factor, $\gamma = 1/(1 - v^2/c^2)^{1/2}$

### Albert Einstein, 1879-1955
The theory of special relativity, 1905 (SR)

## Everything is relative

When talking about the relativity, it is about how one observer perceives things with the help of the information the observers receive with their senses: touch, hearing and sight. To say that it's cold out there can mean for an observer to tremble from the cold, but another observer would say that it is quite comfortable out there. But when we use a thermometer and measure the temperature to -5 degrees then it is -5 degrees. It is a physical measurement. A thermometer does not perceive temperature, it measures the temperature! What the two observers than even are saying about how cold it is out there they have to agree that it is -5 degrees, no more comments!

To perceive events and measuring their coordinates are two different things. Therefore, it feels weird every time I read about the thought experiment with two observers, one of which is still standing on the platform and the other sitting on the train that is moving at a constant speed toward the platform.

In this book, I will describe these thought experiments to define what happens physically, and not how someone observer perceives it one or the other.

SR treats coordinate systems, event, time, place, Lorentz transformations, reference systems, observer, time dilation, thought experiments.

# Events in the coordinate system

An event in spacetime is specified by 4 coordinates. We denote an event with the letter E (from Event). Such an event can be denoted as follows:
 $E = (x, y, z, t)$

To simplify things, we only consider the events taking place on the $x$-axis. Then $y = 0$, $z = 0$ and then we denote the event only by
 $E = (x, t)$.

In these experiments, we will use the material objects that can transmit a light signal and that can record an incoming light signal. Such object on the $x$-axis is a coordinate system. We denote them by S, $S_1$, $S_2$, S' and so on.
We say that they are **material objects** to distinguish them from the light signals which are **wave phenomena**.
The coordinate systems used in our experiments can be stationary to each other or move relative to each other at a constant velocity, $v > 0$.

*The information between these systems is mediated by light signals, moving at the speed of light c. We approximates c to 300 000 km / s.*

# Light

Light, like other electromagnetic radiations, is a wave phenomenon which propagates in space and time. Light moves regardless of how the source or the observer moves.

***But also the direction that the light signal is moving in is independent of how the source or the observer moves.***

***It doesn't matter if the light source moves or rotates, in the moment the signal leaves the light source, the light signal moves with the same speed and the same direction.***

We illustrate how the speed and direction of the light signal are independent of how the light source moves, see Fig. 1.

We consider $S_1$ that transmits a light signal every microsecond, while the source $S_1$ is turning with an arcsecond. In a microsecond the light signal travels a distance of *0.3* km. At a distance of *97,200* km there is the $S_2$. When $S_1$ is facing $S_2$, first light signal is transmitted. After *324,000* microseconds (90x60x60) the light signal reaches $S_2$, and $S_1$ is turned *90* degrees left/right. And it is only the first signal that reaches $S_2$!

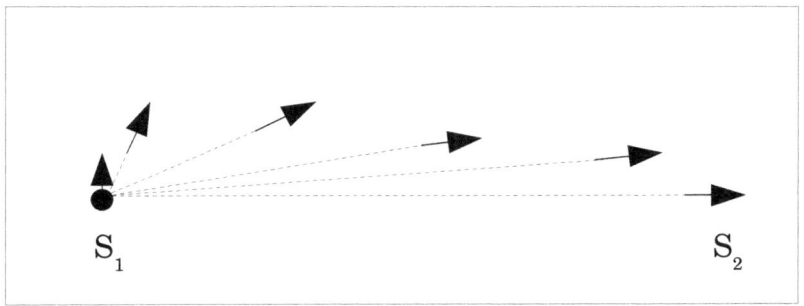

Fig. 1

# Registration, calculation and transformation of coordinates

The theory of special relativity treats two coordinate systems that move relative to each other with constant speed, $v > 0$. The theory of special relativity tell us how to calculate the coordinates of an event in one system using coordinates from the other. Such calculation is called transformation.

Let's look first at a coordinate system and an event, Fig. 2. An event E occurs in the coordinate system $S_1$ at time $t$. $S_1$ get information about the event by registering the light signal from it. **Full details of the event we have only if we know the event's *x*-coordinate, the distance from $S_1$ to E.**

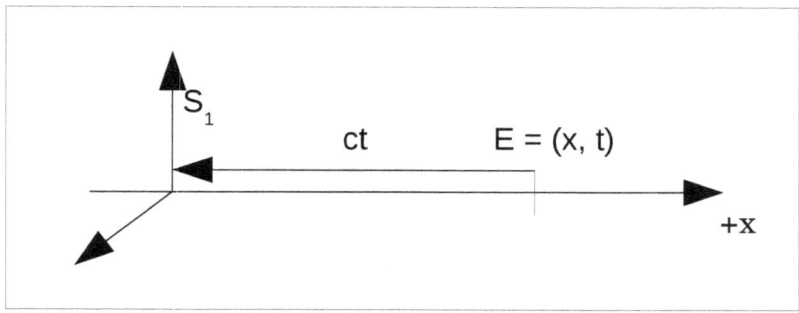

Fig. 2

Then $t = x/c$, and we can express the event with

$$E = (x, x/c)$$

Now we look at two coordinate systems, $S_1$ and $S_2$, which are in rest relative to each other, and an event E. See Fig. 3. Distance between $S_1$ and $S_2$ is $d$.

How does the event $E = (x, t)$ look like when it is registered in $S_1$ and $S_2$?

$E_1 = (x_1, t_1) = (x, x/c)$
$E_2 = (x_2, t_2) = (x-d, (x-d)/c)$

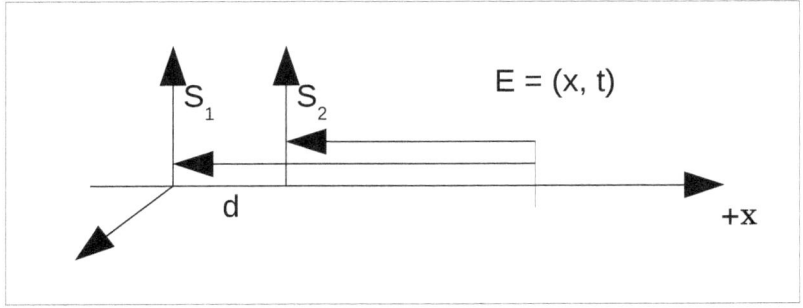

Fig. 3

If we know the distance between $S_1$ and $S_2$, we can calculate the coordinates of the event in one of the coordinates of the event from the other. For example:

$x_2 = x_1 - d$ and $t_2 = t_1 - d/c$

How will it be when $S_2$ moves to the right with a constant velocity $v > 0$? See Fig. 4.
The experiment starts at $t = 0$. $S_1$ and $S_2$ at that time are in the same point, $x_1 = x_2 = 0$.

The only difference between Fig. 3 and Fig. 4 is that instead of distance $d$, we have distance $vt_2$.

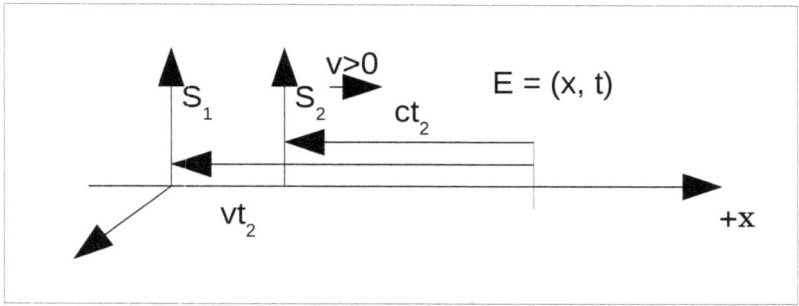

Fig. 4

We have $E_1 = (x_1, t_1) = (x, x/c)$ and using $x_1$ and $t_1$ we calculate $E_2 = (x_2, t_2)$. Here we use the fact that the time it takes for the light signal to reach $S_2$ is the same as the time $S_2$ needs to cover the distance from the point $(0, 0)$ to the point where it meets the light signal. Then we have

$$x = ct_2 + vt_2 \rightarrow t_2 = x/(c+v).$$

$E_2 = (x_2, t_2) = (x_1 c / (c+v), \ t_1 c / (c+v))$

So both $x$- and $t$-coordinate are calculated by the same **factor $c/(c+v)$**.

In the example above, we have placed the event E in **front** of the $S_1/S_2$, if you think about the direction in which $S_2$ is moving.

Now we place the event **behind** $S_1/S_2$, see Fig. 5. In this thought experiment the $x$-coordinates $x, x_1$ and $x_2$ are negative. $t$-coordinates $t, t_1, t_2$ are positive, always.

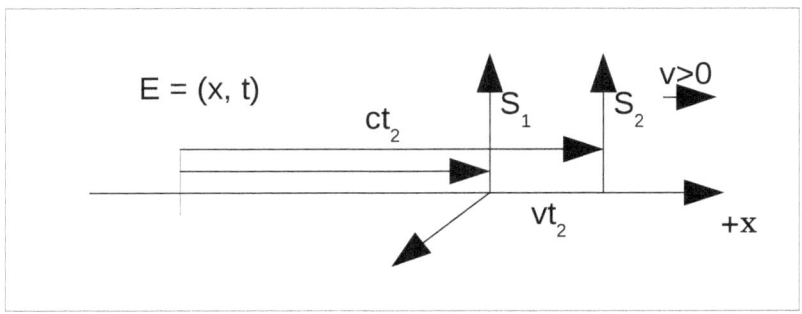

Fig. 5

We have $E_1 = (x_1, t_1) = (x, -x/c)$ and using $x_1$ and $t_1$ we calculate $E_2 = (x_2, t_2)$. Here we use the fact that the time it takes for the light signal to reach $S_2$ is the same as the time $S_2$ needs to cover the distance from the point $(0, 0)$ to the point where the light signal reaches

$S_2$.

This time we have
$$-x = ct_2 - vt_2 \rightarrow t_2 = -x/(c-v)$$

$$E_2 = (x_2, t_2) = (x_1 c/(c-v),\ t_1 c/(c-v))$$

So both $x$- and $t$-coordinate are calculated using the same **factor $c/(c-v)$**.

We can see that the transformation factor is not the same in these two cases, Fig. 4 and Fig. 5, the transformation is **dependent** of where the event happens!

We summarize these two thought experiments to show that the transformation factor between the two inertial reference system is not the same across the $+x$-axis.

We do not need the Lorentz transformations, we will manage the transformation of coordinates between two inertial referens systems by classical physics!

**We have also seen that either the two reference systems are at rest or in motion with respect to each other, the same transformations apply to go from coordinates from one system to another.**

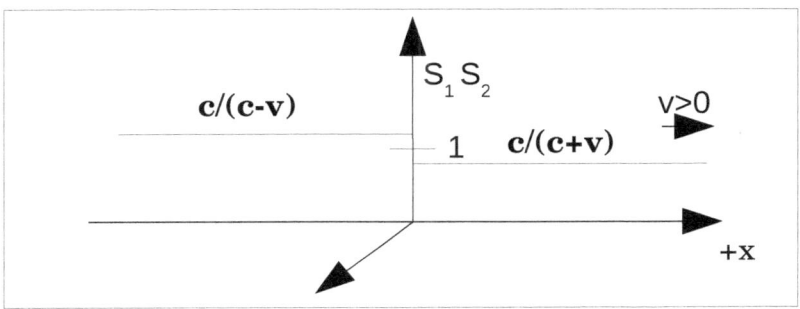

Fig. 6

Above two experiments, Fig. 4 and Fig. 5, and its conclusion should be a thinker for each researcher who works with the special theory of relativity.

This made me conclude that the special theory of relativity contains inaccuracies and that it is therefore incorrect in its entirety!

In the following chapters of this book, I analyze different parts of the special theory of relativity. The analysis shows errors in how to interpret light propagation, how to derive Lorentz transformations.

It's about basic physics and mathematics!

# Analysis of time dilation

In some of the literature [1], [6], [8], which deals with the theory of special relativity, time dilation is explained as follows, and one uses the same thought experiment to derive the Lorentz factor.

In these thought experiments one uses spaceship in which a light beam starts on the floor, is reflected in the ceiling and comes back to the floor. We illustrate the two cases.

The first case is when the spacecraft is stationary, see Fig. 7. Distance from the floor to the ceiling is $L$. Then the time that the light need to cover the distance floor-ceiling-floor is

$$t_0 = 2L/c$$

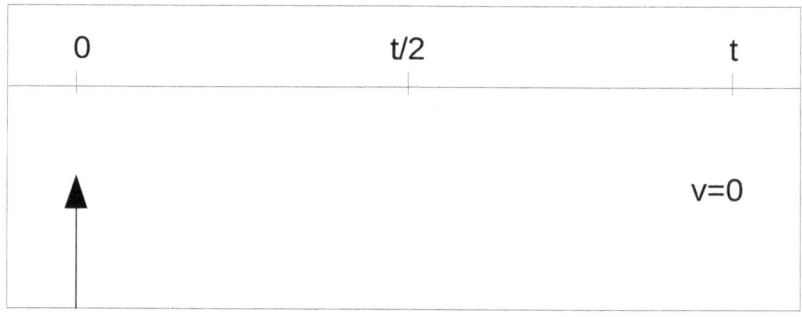

Fig. 7

The second case is when the spacecraft is moving with constant speed $v > 0$ to the right, Fig. 8.

We consider the triangle with given sides and calculate from there:

$t = 2L/(c^2-v^2)^{1/2}$

One replaces *2L* with $t_0 c$ and we get

$t = t_0 c/(c^2-v^2)^{1/2} = t_0 \gamma$ where $\gamma$ is Lorentz factorn.

**This is what the theory of special relativity says.**

Here, I think that Fig. 8 is the most absurd, dissolved from reality, explanation of a physical phenomenon I've seen so far!

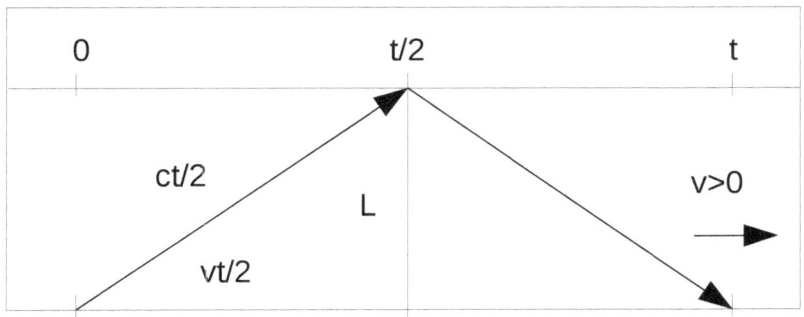

Fig. 8

## My explanation:
***A light beam moves at a constant speed c and with the same direction regardless of how the light source and the observer moves.***

Imagine a stationary platform in the vacuum of space. A light signal leaves the platform and will move with the same direction, see Fig. 9.

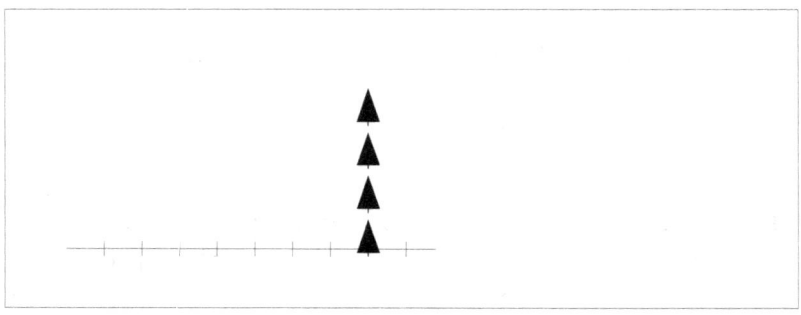

Fig. 9

Consider now the same platform in vacuum, in space, Fig. 10, moving at speed $v > 0$ to the right. A light beam leaves the platform and will move in the same direction.

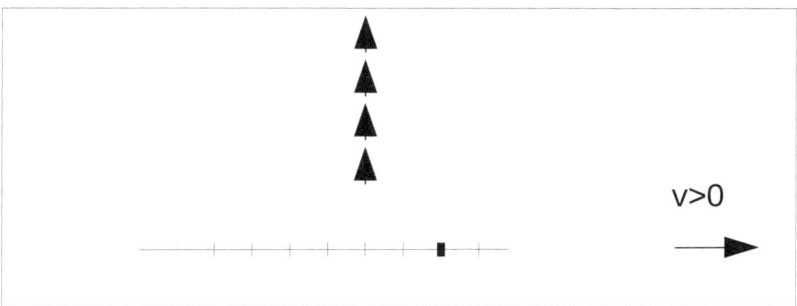

Fig. 10

We illustrate the reasoning that a light signal leaving the floor, reflecting in the ceiling and hits the floor again, is moving in the same direction.
We will, in the same image, Fig. 11, show more intermediate positions so we in a simple way can see how the light signal and the "spaceship" moves.

We have a "spaceship" moving with constant velocity $v = 30 \ km/s$ to the right. We consider a light signal is leaving the floor, reflects itself in the ceiling and reaches the floor again. During this time the ship moves with a distance $d = 2x$.

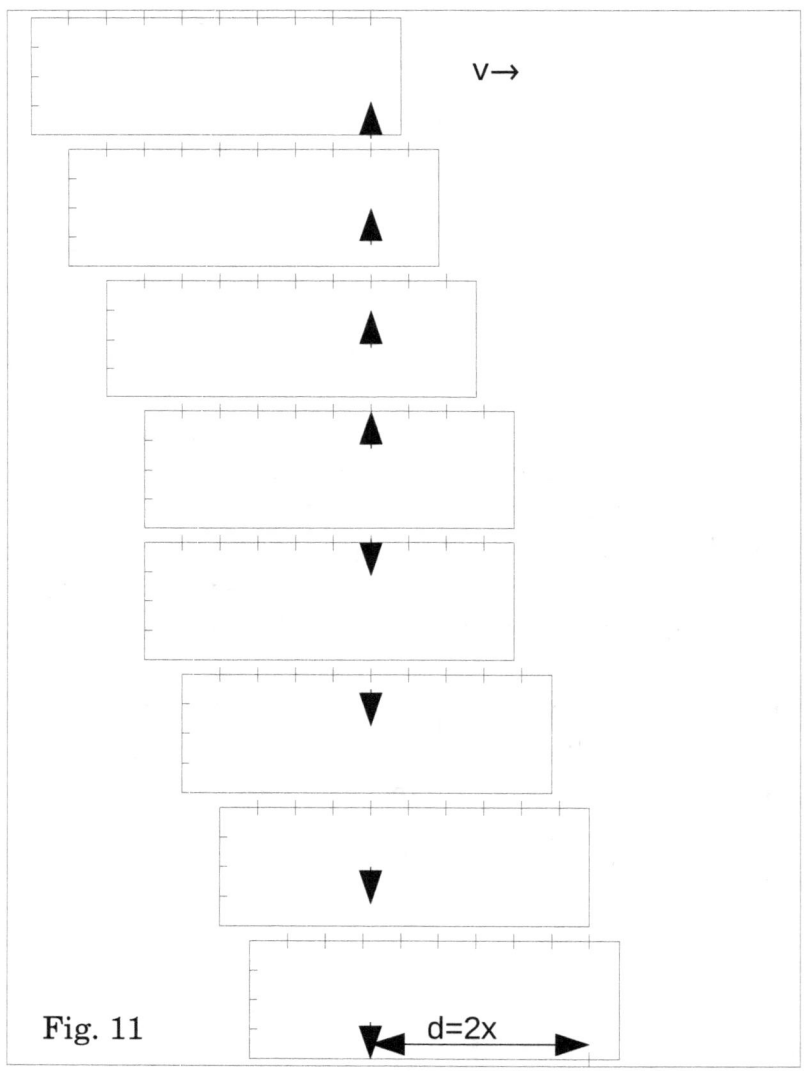

Fig. 11

Carefully consider the picture! A light signal starts from the floor, reflects itself in the ceiling and end up in another point on the floor, **behind** the point from which it started, if you think about the direction of movement.

## The light propagates not in zigzag.

Distance between two points only tell us how far the ship moved in the same time as the light signal traveled distance $2L$. We denote this distance $d = 2x$.

We summarize this: The time during which the light signal had passed distance $2L$ is the same as the spaceship needs to cover the distance $2x$.

$t = 2L/c = 2x/v \rightarrow x = Lv/c$

Example: $L = 10\ m,\ v = 30\ km/s,\ c = 300\ 000\ km/s$

$x = 10*30/300000\ m = 1/1000\ m = 1\ mm$

*This means that you could build a device that would measure the speed of the Earth in space, around the Sun, around the galactic center.*

$v = xc/L$

This device would work as an electromagnetic gyroscope, a **light gyroscope**.

**Note that the time which the light signal travels distance 2L is the same, either the system is at rest or is moving at a constant speed $v > 0$!**

**Then we have no time dilation!**

Below I present six different calculations showing that the derivation of Lorentz transformations is incorrect.

My motto:
*When we study the physical phenomena, we always make a mathematical model of them. In such a model, there are built-in physical laws that are held together by mathematical tools. If the description of the physical phenomenon is correct, the mathematical model has no errors!*

The theory of special relativity deals with the relationship between two inertial reference systems, S and S', moving towards each other with constant velocity $v > 0$. Every event in the reference system is determined by four coordinates, three for space and one for time. To determine the event's coordinates in one of the reference system with the help of the event's coordinates in the second one uses Lorentz transformations:

$E = (x, y, z, t)$, an event in S
$E' = (x', y', z', t')$, an event in S'

To facilitate understanding of the calculations one usually puts $y = y'$ and $z = z'$. Then the Lorentz transformations become:

$x' = (x - vt)\gamma$                  (LT$_1$)
$t' = (t - vx/c^2)\gamma$            (LT$_2$)

where $\gamma = 1/(1 - v^2/c^2)^{1/2}$ is called the Lorentz factor.

# Derivation of Lorentz transformations
# Example 1

In the special theory of relativity, Lorentz transformations are used to calculate the coordinates of the events in a reference system using coordinates in another reference system which moves towards each other at constant speed, **$v > 0$**.

We follow the reasoning and the calculations from *[7], pages 14-15*. Here one uses the following:

(2-3) $\quad x' = u't'$ and $x = ut$
(2-4) $\quad x' = Ax+Bt, \quad t' = Cx+Dt$

It is said that between *(x, t)* and *(x', t')* there must be a linear transformation. This in turn means that A, B, C, D are constants.
To determine above four constants, three special cases are used.

c1) The object in which event E occurs is in the origin of $S_2$.
$\quad E_2 = (x_2, t_2) = (0, t)$
c2) The object in which event E occurs is in the origin of $S_1$.
$\quad E_1 = (x_1, t_1) = (0, t)$

c3) One equates the object in which event E occurs with a light beam.

We follow the calculations:

c1) $x' = 0$, $x = vt$
We replace these in (2-4) and get:

$0 = Avt + Bt$ and $t' = Cvt + Dt \rightarrow$
**$B = -Av$ and $t' = Cvt + Dt$**

c2) $x = 0$, $x' = -vt'$
We replace these in (2-4) and get:

$-vt' = Bt$ and $t' = Dt$
You divide these two equations and get

**$B = -Dv \rightarrow D = A$**

## My addition:
But $t' = Cvt + Dt$ from c1 and $t' = Dt$ from c2 $\rightarrow$
$Cvt + Dt = Dt \rightarrow Cvt = 0 \rightarrow$ **$C = 0$**

Can you combine c1 and c2?
Then it becomes
(2-4)   $x' = Ax - Avt$ and $t' = At$ or
**$x' = A(x-vt)$ and $t' = At$**

Now we use
c3) $x' = ct'$ and $x = ct$
We replace these in $x' = A(x-vt)$ and get

$ct' = A(ct-vt)$ and $t' = At$

From here we get:

$ct = ct-vt \rightarrow vt = 0 \rightarrow v = 0$ or $t = 0$

If $t = 0$ then $S_1$, $S_2$ at the same point, no Lorentz transformations are needed. If $v = 0$ then we have a contradiction with origin conditions.

# Example 2

This derivation is in *[7], pages 14-15*.

The derivation of Lorentz transformations is made with the assumption that these transformations must be linear:

$x' = Ax + Bt$
$y' = Cx + Dt$, where $A$, $B$, $C$ and $D$ are constants.

Three special cases are used to solve this equation system:

c1) $x' = 0$, $x = vt$
c2) $x = 0$, $x' = -vt'$
c3) $x = ct$ and $x' = ct'$, where $c$ is the speed of light

and finally you come to Lorentz transformations

$$x' = (x - vt)\gamma \qquad (LT_1)$$
$$t' = (t - vx/c^2)\gamma \qquad (LT_2)$$

where $\gamma = 1/(1 - v^2/c^2)^{1/2}$ is called the Lorentz factor.

However, if Lorentz transformations $LT_1, LT_2$ have been derived using c1, c2 and c3, these three special cases should verify Lorentz transformations $LT_1, LT_2$ without obtaining mathematical contradiction.

## My evidence:
From c1 and $LT_1$ → $0 = (vt - vt)\gamma$ → $0 = 0$, OK
From c1 and $LT_2$ → $t' = (t - v(vt)/c^2)\gamma$ →
$t' = t(1-v^2/c^2)\gamma$

From c2 and $LT_1$ → $-vt' = (0-vt)\gamma$ → $-vt' = -vt\gamma$ →
$t' = t\gamma$
From c2 and $LT_2$ → $t' = (t-v0/c^2)\gamma$ → $t' = t\gamma$, same result

But the result from c1 and $LT_2$ is $t' = t(1-v^2/c^2)\gamma$

and the result from c2 and $LT_2$ is $t' = t\gamma$

$$\rightarrow 1-v^2/c^2 = 1 \rightarrow v = 0$$

**This result, $v = 0$, is in contradiction to the theory's assumption that the two reference systems move toward each other at constant velocity $v > 0$!**

This shows that the special theory of relativity contains errors.

Can you combine c1 and c2? If not then you wonder why you get so different results? Why we get so different relationships between $t'$ and $t$?

# Example 3

Below we follow *[3], page 125; Appendix;*
*A simple derivation of the Lorentz transformation*

In his presentation of the special theory of relativity, Einstein finally comes to Lorentz transformations:

$$x' = (x - vt)\gamma \qquad (LT_1)$$
$$t' = (t - vx/c^2)\gamma \qquad (LT_2)$$

where $\gamma = 1/(1 - v^2/c^2)^{1/2}$ is called the Lorentz factor.

I will quote Einstein and analyze what he claims:

Einstein:
"A light signal that runs along the positive x-axis is propagated according to the equation

$$x = ct \text{ or } x\text{-}ct = 0 \text{ "} \qquad (1)$$

The expression "along the positive x-axis" means that equations (1) apply to $x \geq 0$.
Similar applies to the other coordinate system.

$$x' = ct' \text{ or } x'\text{-}ct' = 0 \qquad (2)$$

Equations (2) apply for $x' \geq 0$.

Einstein:
"The space times events that meet (1) must also meet (2). This is obviously the case if we generally have the relationship

$$(x'\text{-}ct') = \lambda(x\text{-}ct) \qquad (3)$$

where $\lambda$ is a constant. For according to (3) becomes $x'\text{-}ct'$ equal to zero if $x\text{-}ct$ equals zero."

Einstein:
"An analogous view of a beam of light propagating along the negative x-axis gives the condition

$$(x'+ct') = \mu(x+ct)"  \quad (4)$$

This section applies to $x \leq 0$ and $x' \leq 0$.

Equation (3) applies for **$x \geq 0$** and for **$x' \geq 0$.**
Equation (4) applies for **$x \leq 0$** and for **$x' \leq 0$.**

Einstein:
"If you now add respective subtract the equations (3) and (4), you get:

$x' = ax - bct$
$ct' = act - bx$"

and so on...
Furthermore, we do not need to analyze Einstein's derivation of Lorentz transformations.

**Here, Einstein makes a basic mathematical error: one adds and subtracts equations that apply in completely different validity areas.**
I quote *[10], page 32:*
"If $f$ and $g$ are functions, then for each $x$ that belongs to

validity areas for both $f$ and $g$, we define function $f + g$ ..."

*We can do operations on functions **only** in their common areas of validity.*
Above areas, equations (3) and (4), have a single point in common:

$x = 0, x' = 0$.

But then from (1) → $t = 0$ and from (2) → $t' = 0$ and then we have the trivial example when both coordinate systems are in the same point!

Then we cannot talk about two reference systems that move at constant speed $v > 0$ to each other! Then no transformations are needed to move from one to the other! They are identical. Then no theory is needed that deals with the relationship between these two coordinate systems!

# Example 4

Below we follow *[3], page 125; Appendix;*
*A simple derivation of the Lorentz transformation*
In his presentation of the special theory of relativity, Einstein finally comes to Lorentz transformations:

$$x' = (x - vt)\gamma \qquad (\text{LT}_1)$$
$$t' = (t - vx/c^2)\gamma \qquad (\text{LT}_2)$$

where $\gamma = 1/(1 - v^2/c^2)^{1/2}$ is called the Lorentz factor. I will quote Einstein and analyze what he claims:

Einstein:
"A light signal that runs along the positive x-axis is propagated according to the equation"

$$x = ct \text{ or } x\text{-}ct = 0 \qquad (1)$$

Similar applies to the other coordinate system.

$$x' = ct' \text{ or } x'\text{-}ct' = 0 \qquad (2)$$

Einstein:
"The space times events that meet (1) must also meet (2). This is obviously the case if we generally have the relationship

$$(x'\text{-}ct') = \lambda(x\text{-}ct) \qquad (3)$$

where $\lambda$ is a constant, because according to (3) becomes $x'\text{-}ct'$ equal to zero if $x\text{-}ct$ equals zero ". Here you have to state that

$$\lambda \mathrel{!}= 0 \qquad (3.1)$$

For if $\lambda = 0$ one cannot say that "$x'$-$ct$" equals zero if $x$-$ct$ equals zero".
Because if $\lambda = 0$ then $x'$-$ct'$ = 0 even if $x$-$ct$ != 0.

Einstein:
"An analogous view of a beam of light propagating along the negative x-axis gives the condition":

$$(x'+ct') = \mu(x+ct) \qquad (4)$$

Here, too, one must state that
$$\mu \;!=\; 0 \qquad (4.1)$$
Below I use $A$, $B$ instead of $a$, $b$:

Einstein:
"If you now add respectively subtract the equations (3) and (4), you get:

$$x' = Ax - Bct \qquad (5.1)$$
$$ct' = Act - Bx \qquad (5.2)$$

where we are introduced for convenience $A = (\lambda+\mu)/2$ and $B = (\lambda-\mu)/2$.
Now our task would be solved if we knew the constants $A$ and $B$. We find them by the following considerations:"
Furthermore, Einstein uses the following three

conditions:

c1) $x' = 0$
c2) $t = 0$
c3) $t' = 0$

From c1) and (5.1) $\to x = ctB/A$
From c3) and (5.2) $\to x = ctA/B$
$\to B/A = A/B \to A \mathrel{!=} 0$ and $B \mathrel{!=} 0$ and $A^2 = B^2$
$\to ((\lambda+\mu)/2)^2 = ((\lambda-\mu)/2)^2 \to (\lambda+\mu) = \pm(\lambda-\mu)$
$\to \lambda+\mu = \lambda-\mu \to 2\mu = 0 \to \mu = 0$ contradicts (4.1) or
$\to \lambda+\mu = -\lambda+\mu \to 2\lambda = 0 \to \lambda = 0$ contradicts (3.1).
This means that the derivation of Lorentz transformations is incorrect!

# Example 5

Below we follow *[3], page 125; Appendix;*
*A simple derivation of the Lorentz transformation*

In his presentation of the special theory of relativity, Einstein finally comes to Lorentz transformations:

$x' = (x - vt)\gamma$              (LT$_1$)
$t' = (t - vx/c^2)\gamma$           (LT$_2$)

where $\gamma = 1/(1 - v^2/c^2)^{1/2}$ is called the Lorentz factor, $c$ is the speed of light.

This factor is: $\gamma > 1$ $(v > 0)$, $\gamma < +\infty$ $(v < c)$.

Einstein:
"A light signal that runs along the positive x-axis is propagated according to the equation"

$$x = ct \text{ or } x\text{-}ct = 0 \qquad (1)$$

Similar applies to the other coordinate system.

$$x' = ct' \text{ or } x'\text{-}ct' = 0 \qquad (2)$$

Einstein:
"The space times (events) that meet (1) must also meet (2). This is obviously the case if we generally have the relationship

$$(x'\text{-}ct') = \lambda(x\text{-}ct) \qquad (3)$$

where $\lambda$ is a constant, because according to (3), $x'\text{-}ct'$ equals zero if $x\text{-}ct$ equals zero.

Einstein:
"An analogous view of a beam of light propagating along the negative x-axis gives the condition":
$$(x'+ct') = \mu(x+ct) \qquad (4)$$
Einstein:
"If you now add respectively subtract the equations (3)

and (4), you get:

$$x' = Ax - Bct \qquad (5.1)$$
$$ct' = Act - Bx \qquad (5.2)$$

where we are introduced for convenience

$$A = (\lambda + \mu)/2 \text{ and } B = (\lambda - \mu)/2.$$

Now our task would be solved if we knew the constants $A$ and $B$. We find them by the following considerations:"

Furthermore, Einstein uses the following three conditions:

c1) $x' = 0$
c2) $t = 0$
c3) $t' = 0$

## Mathematical examination:
$LT_1$, c1 $\to$ $x' = 0$, $x = vt$
$LT_2$, c1 $\to$ $x' = 0$, $t' = (t - vx/c^2)\gamma$
$LT_1$, c2 $\to$ $t = 0$, $x' = x\gamma$
$LT_2$, c2 $\to$ $t = 0$, $t' = -vx\gamma/c^2$
$LT_1$, c3 $\to$ $t' = 0$, $x' = (x - vt)\gamma$
$LT_2$, c3 $\to$ $t' = 0$, $t = vx/c^2$

All these results should verify $LT_1$ and $LT_1$ because conditions c1, c2 and c3, **all** were used in the derivation of $LT_1$ and $LT_2$.

We take $LT_1$, c1 and $LT_2$, c3:
$x = vt$ and $t = vx/c^2 \to x = vvx/c^2 \to 1 = v^2/c^2 \to v^2 = c^2$
$\to v = \pm c!$

This means that the derivation of Lorentz transformations is incorrect!

# Example 6

Below we follow *[3], page 125; Appendix;*
*A simple derivation of the Lorentz transformation*

In his presentation of the special theory of relativity, Einstein finally comes to Lorentz transformations:

$x' = (x - vt)\gamma$                  $(LT_1)$
$t' = (t - vx/c^2)\gamma$            $(LT_2)$

where $\gamma = 1/(1 - v^2/c^2)^{1/2}$ is called the Lorentz factor, $c$ is the speed of light.
This factor is: $\gamma > 1$ $(v > 0)$, $\gamma < +\infty (v < c)$.
Einstein:
"A light signal that runs along the positive x-axis is propagated according to the equation"

$x = ct$ or $x-ct = 0$ \hfill (1)

Similar applies to the other coordinate system.

$x' = ct'$ or $x'-ct' = 0$ \hfill (2)

Einstein:
"The space times (events) that meet (1) must also meet (2). This is obviously the case if we generally have the relationship

$(x'-ct') = \lambda(x-ct)$ \hfill (3)

where $\lambda$ is a constant, because according to (3), $x'-ct'$ equals zero if $x-ct$ equals zero".
**Here you should specify that $\lambda \ != 0$.**

Einstein:
"An analogous view of a beam of light propagating along the negative x-axis gives the condition":

$(x'+ct') = \mu(x+ct)$ \hfill (4)

**Here you should specify that $\mu \ != 0$.**

Einstein:
"If you now add respectively subtract the equations (3)

and (4), you get:

$$x' = Ax - Bct \qquad (e1)$$
$$ct' = -Bx + Act \qquad (e2)$$

where we are introduced for convenience

$$A = (\lambda + \mu)/2 \text{ and } B = (\lambda - \mu)/2.$$

Now our task would be solved if we knew the constants $A$ and $B$. We find them by the following considerations:"

Furthermore, Einstein uses the following three conditions:

c1) $x' = 0$
c2) $t = 0$
c3) $t' = 0$

## Mathematical examination:
e1, c1 → $0 = Ax - Bct$ → $Ax = Bct$ → $x = (B/A)ct$
e2, c1 → $ct' = -Bx + Act$
e1, c2 → $x' = Ax$
e2, c2 → $ct' = -Bx$
e1, c3 → $x' = Ax - Bct$
e2, c3 → $0 = -Bx + Act$ → $Bx = Act$ → $x = (A/B)ct$

We have received:

r1) $x = (B/A)ct$
r2) $ct' = -Bx+Act$
r3) $x' = Ax$
r4) $ct' = -Bx$
r5) $x' = Ax-Bct$
r6) $x = (A/B)ct$

We combine and receive:
r1, r6 $\rightarrow A/B = B/A \rightarrow A \mathrel{!}= 0$ and $B \mathrel{!}= 0$
r3, r5 $\rightarrow Ax = Ax - Bct \rightarrow -Bct = 0 \rightarrow Bt = 0 \rightarrow t = 0$
r2, r4 $\rightarrow -Bx = -Bx + Act \rightarrow Act = 0 \rightarrow At = 0 \rightarrow t = 0$
$\rightarrow x = 0, t = 0, x' = 0, t' = 0$

Then we have the trivial case, when both coordinate systems are in the same point! There is no need for transformation of coordinates from S to S', then no theory is needed for this!

## Michelson-Morley experiment, 1887

Much has been written about this experiment. It is described as one of the most important and famous experiments in the history of physics.

But the result was not as expected.
It was so-called negative result. And this ultimately leads to the special theory of relativity.

In my analysis of this experiment, I show why it was so. I will show pictures of four different positions of the interferometer, show how the light rays move, make calculations on their passed distances.

In my analysis, I apply the principle that **the light moves independently of the source and the observer's motion**.

The pictures are not drawn to scale and I present only the most necessary elements to simplify as much as possible, so that one can see the most important.

In the first image, Fig. 12, we show the interferometer when the light beam is transmitted in the same direction as the apparatus moves.

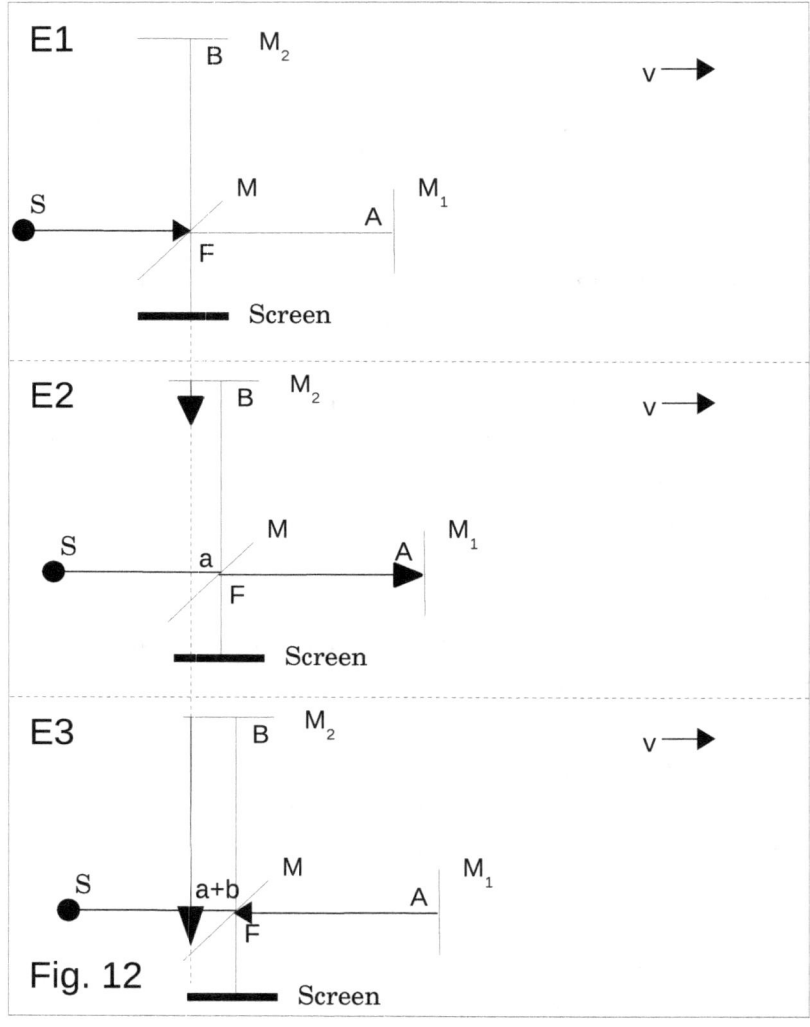

Fig. 12

**E1:**
Interferometers arms: FA = FB = L (*length*).
A light beam is transmitted from S and is divided into two in F. The light beam $S_1$ continues straight ahead towards A (mirror $M_1$). $S_2$ is reflected at an angle of 90° and goes to B (mirror $M_2$).

**E2:**
When $S_1$ reaches A, it is reflected and goes back to F.
When $S_2$ reaches B, it is reflected and goes back to M.

While $S_1$ approaches A and reaches this point, the entire system moves with the distance $a$. During this time, $S_2$ pass the same distance $L+a$. $a/v = (L+a)/c$

**E3:**
While $S_1$ goes towards F and reaches this point, the entire system moves with the distance $b$. Then the $S_1$ pass the distance $L-b$. During this time, $S_2$ pass the same distance $L-b$. $b/v = (L-b)/c$

We now calculate the length of the distances that S1 and S2 pass.

$a = Lv/(c-v)$
$b = Lv/(c+v)$
$a+b = 2Lcv/(c^2-v^2)$
$a-b = 2Lv^2/(c^2-v^2)$

$$\text{Length}(S_1) = \text{Length}(S_2) = 2L + (a - b)$$

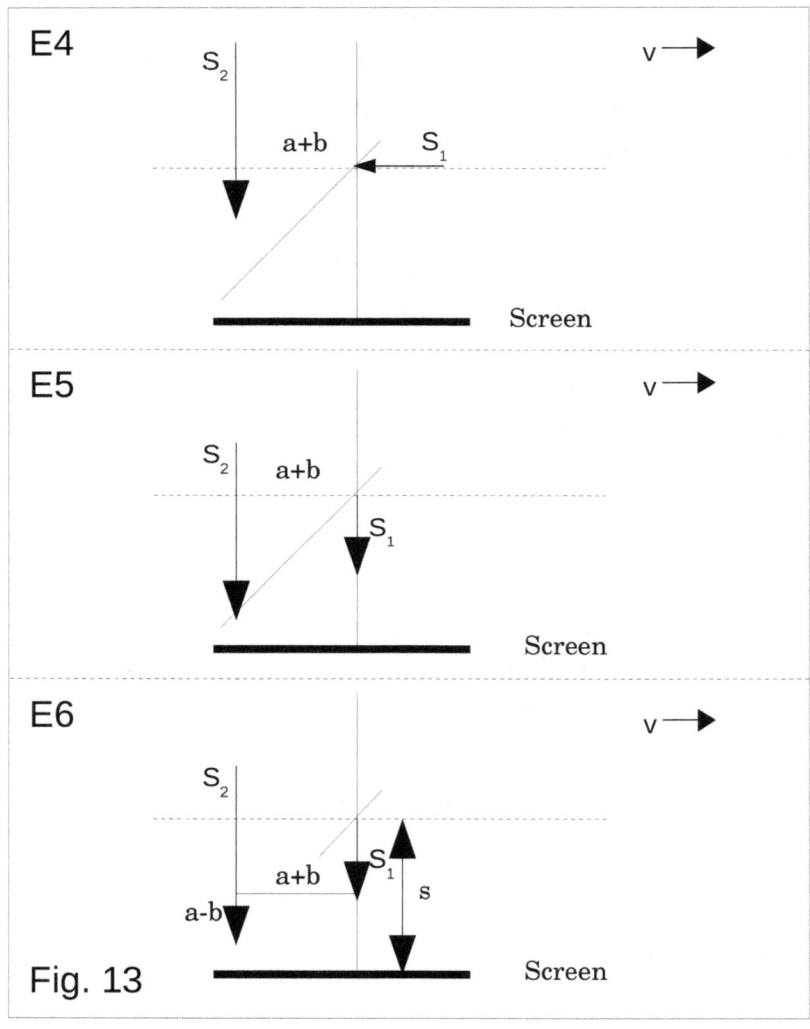

Fig. 13

In the figure Fig. 13 we show the moment when $S_1$ reaches the point F (mirror M) for the second time and then continues towards the screen.

Then the spatial distance between the two light rays becomes equal to $a + b$.
The light beam $S_2$ reaches earlier the screen, time difference then becomes $\Delta t = (a-b)/c$.

If we take the distance between the mirror M and the screen equal to $s$, the parameters of the interference pattern can be calculated. $y = s\lambda/d$

The size of the distance between two wave peaks of the interference pattern depends only on the distance $s$ between the mirror M (point F) and the screen and the distance $d = a + b$ between the two light beams.

See picture E6 in Fig. 13.

We are now turning the device 90° counterclockwise. See Fig. 14.

We also make similar calculations here.

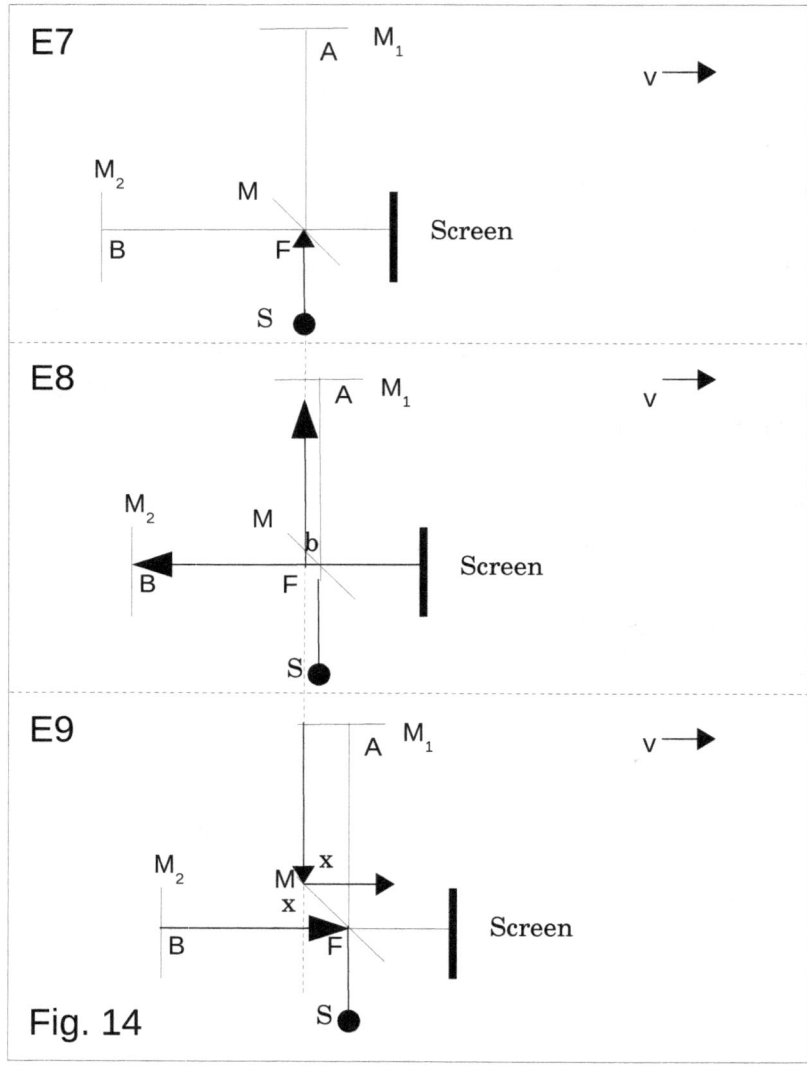

Fig. 14

E7:
Interferometers arms: FA = FB = L.
A light beam is transmitted from S and is divided into two in F. The light beam $S_1$ continues straight ahead towards A (mirror $M_1$). $S_2$ goes to B (mirror $M_2$).

E8:
When $S_2$ reaches B, it is reflected and goes back to F. When $S_1$ reaches $M_1$, it is reflected and goes back to M.

As $S_2$ approaches B and reaches this point, the entire system moves with distance $b$. Then the $S_2$ pass the distance $L$-$b$. During this time, $S_1$ pass the same distance $L$-$b$.

E9:
While $S_2$ goes towards F and reaches this point, the entire system moves with $a$. Then the $S_2$ pass the distance $L + a$. During this time, $S_1$ pass the same distance $L + a$.

But we must calculate the moment when $S_1$ reaches the mirror M for the second time.
We analyze this in Fig. 15.

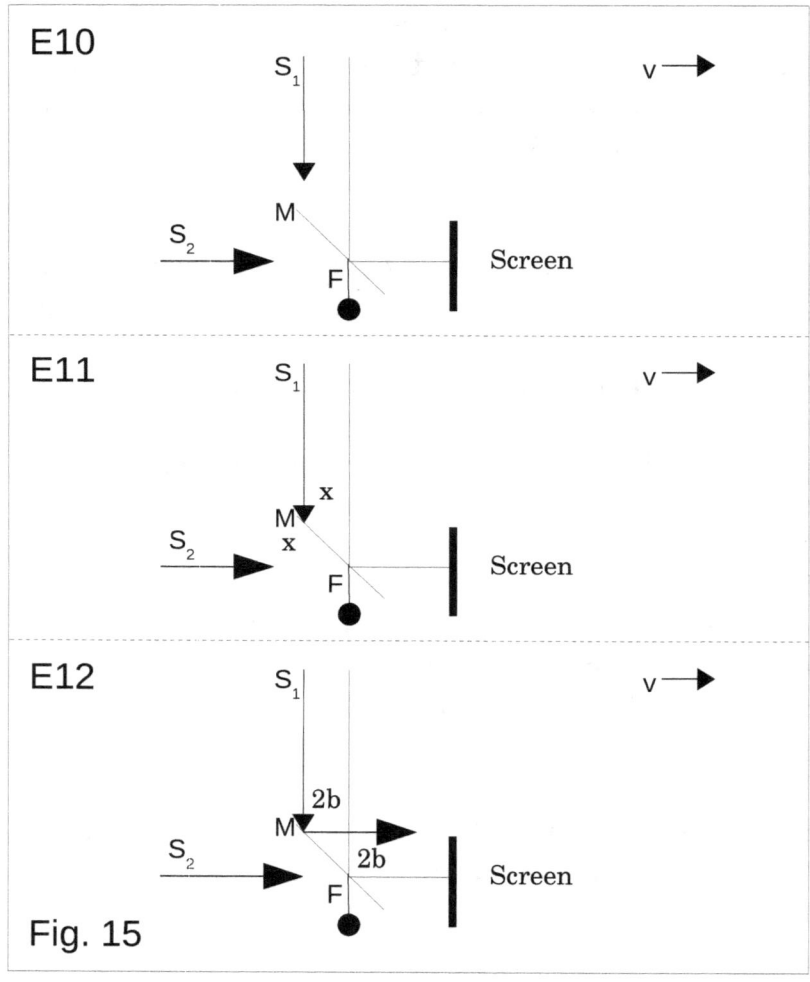

Fig. 15

The calculation:

We say that the device pass the distance $x$ at the moment the light beam $S_1$ reaches the mirror M for the second time.
The time for this is then $x/v$.

During this time, the light beam $S_1$ has traveled the distance $2L-x$.
Then we have:

$x/v = (2L-x)/c$
$x/v = 2L/c - x/c$
$x/v + x/c = 2L/c$
$x(c+v)/cv = 2L/c$
$x = 2Lv/(c+v)$

$x = 2b$

This means that the spatial distance between the two light beams is $d = 2b$.

We go on and turn the interferometer with 90° counterclockwise.

We show how the light rays move in Fig. 16.

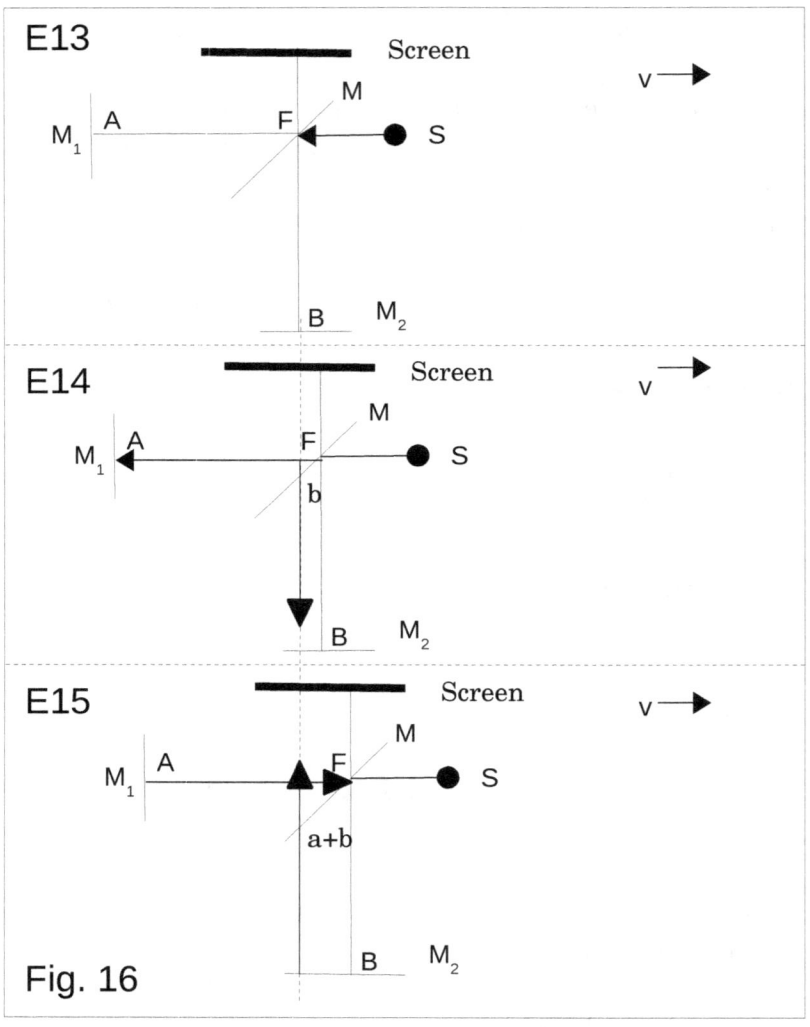

Fig. 16

E13:

Interferometers arms: FA = FB = L.
A light beam is transmitted from S and is divided into two in F. The light beam $S_1$ continues straight ahead towards A (mirror $M_1$). $S_2$ goes to B (mirror $M_2$).

E14:
When $S_1$ reaches A, it is reflected and goes back to F.
When $S_2$ reaches B, it is reflected and goes back to M.

While $S_1$ approaches A and reaches this point, the entire system moves by distance $b$. Then the $S_1$ pass the distance $L-b$. During this time, $S_2$ pass the same distance $L-b$.

E15:
While $S_1$ goes towards F and reaches this point, the whole system moves with $a$, $S_1$ pass the distance $L + a$. During this time, $S_2$ pass the same distance $L + a$.

This part of the experiment is similar to that of Fig. 12 in that the spatial distance between the two light beams when reaching the screen also becomes $d = a + b$.

Last part of the experiment. We turn the interferometer with another 90° counterclockwise. Now we have turned the interferometer with a full

270°. See picture Fig. 17.

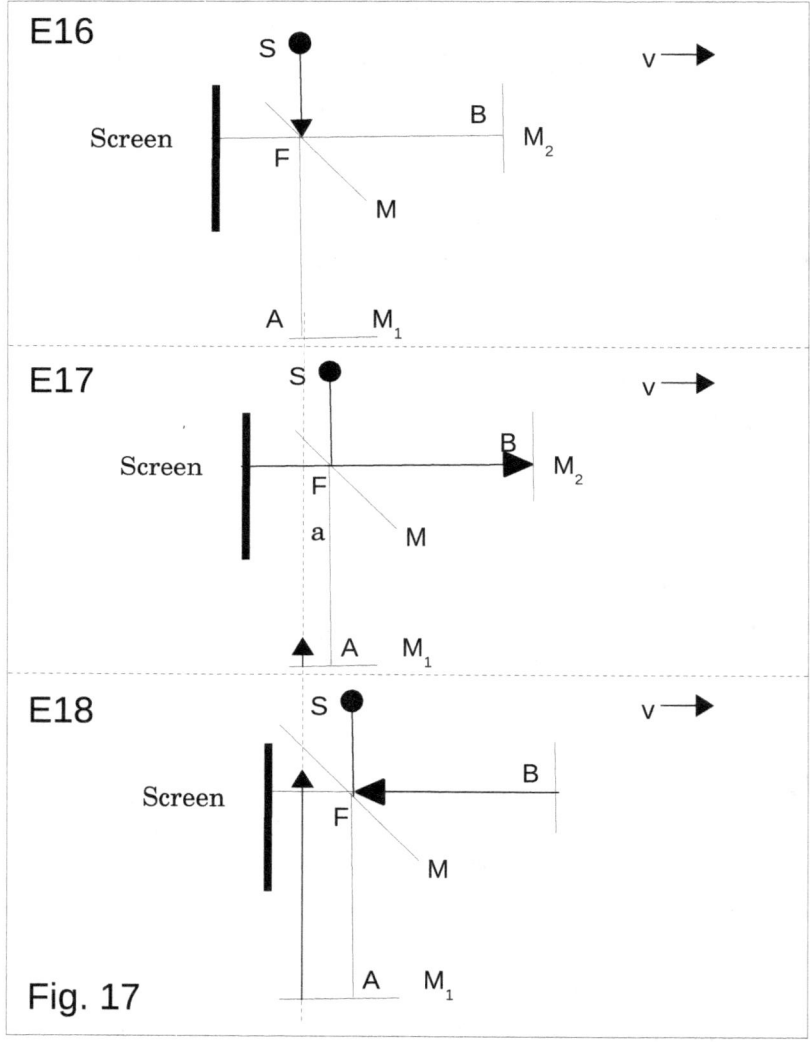

Fig. 17

E16:
Interferometers arms: $FA = FB = L$.
A light beam is transmitted from S and is divided into two in F. The light beam $S_1$ continues straight ahead towards A (mirror $M_1$). $S_2$ goes to B (mirror $M_2$).

E17:
When $S_2$ reaches B, it is reflected and goes back to F. When $S_1$ reaches $M_1$, it is reflected and goes back to M.

While $S_2$ approaches B and reaches this point, the entire system moves by distance $a$. Then the $S_2$ pass the distance $L + a$. During this time, $S_1$ pass the same distance $L + a$.

E18:
While $S_2$ goes towards F and reaches this point, the whole system moves with $b$. Then the $S_2$ pass the distance $L\text{-}b$. During this time, $S_1$ pass the same distance $L\text{-}b$.

But we must calculate the moment when $S_1$ reaches the mirror M for the second time.

We analyze this in Fig. 18.

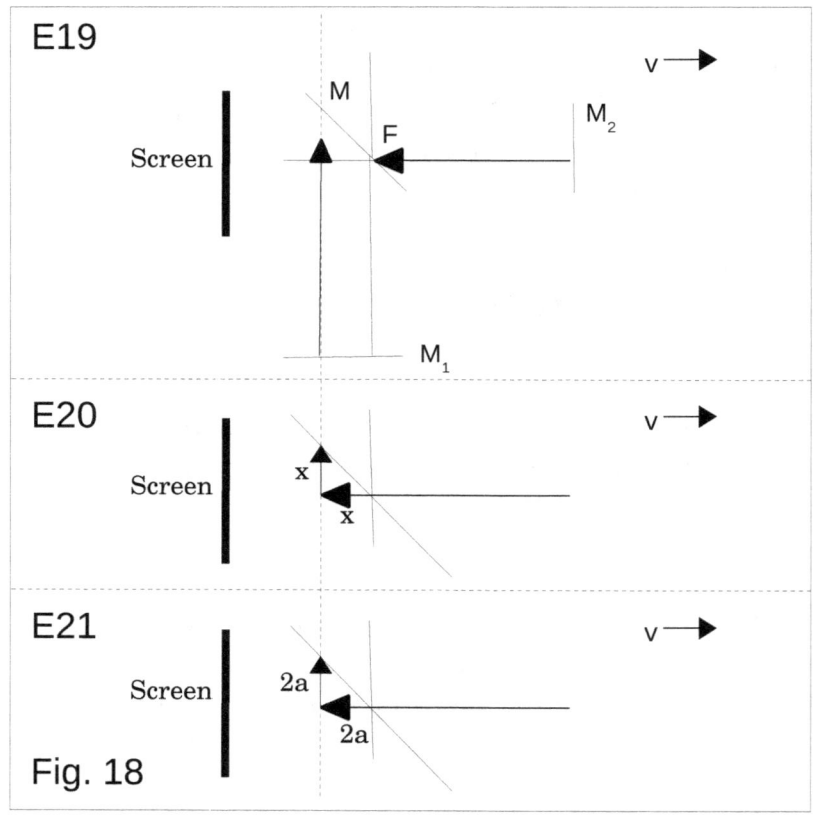

Fig. 18

When $S_1$ reaches mirror M for the second time, it has passed a distance of $2L + x$. The time then becomes $(2L + x)/c$ and is the same time as $x/v$.

$x/v = (2L+x)/c$
$x = 2a$

Then the spatial distance between the two light rays becomes equal to $2a$.

The spatial distance d between the two light beams reaching the screen is as follows:

Fig. 12-13   Exp. 1   $d = a+b$
Fig. 14-15   Exp. 2   $d = 2b$
Fig. 16      Exp. 3   $d = a+b$
Fig. 17-18   Exp. 4   $d = 2a$

We see that $d$ reaches its maximum/minimum when the light beam is sent exactly at the right angle to the direction of movement of the apparatus. See Fig. 18A.

Should we turn the middle mirror M by 180° then $2b$ would change place with $2a$. And it is between these two part experiments when the interference pattern reaches the biggest difference!

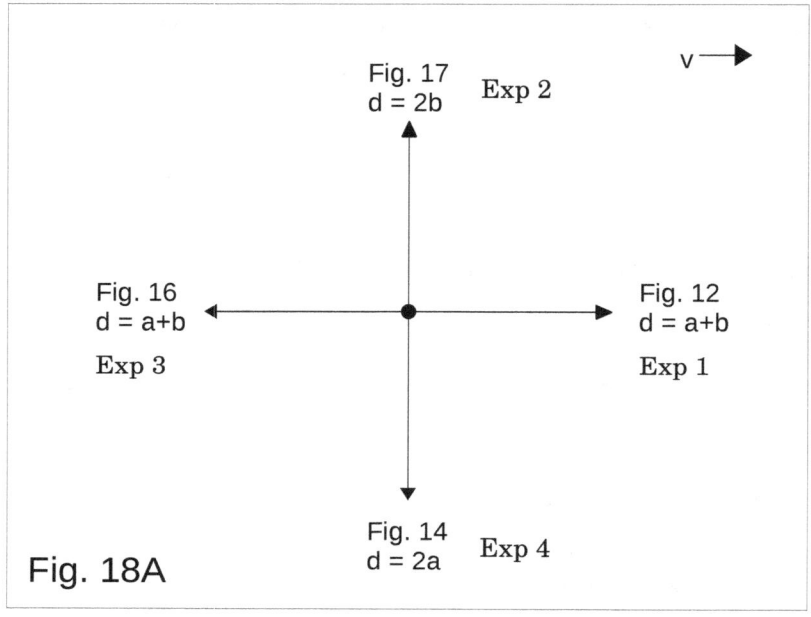

Fig. 18A

But not even this is the whole truth! This applies if we direct the interferometer either along the same direction as the reference system moves or at right angles to it.

**<u>When working with the light, it is extremely important to remember that you do NOT know the absolute speed of a reference system, nor its direction!</u>**

The calculations we made before, we should redo and take into account that both interferometer arms can move sideways. See Fig. 19.

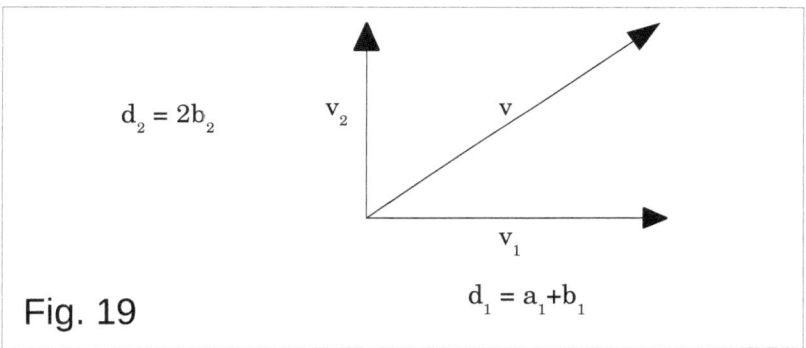

Fig. 19

Why? We do not know the exact direction of Earth's velocity nor its value.
We do not know the interferometer's absolute motion!

Because:
- The Earth has a movement around its own axis
- Earth moves around the sun
- The solar system moves around the center of the galaxy
- The galaxy moves towards the Great Attractor.
- And ...

What is the result of all these speeds? How then does the interference pattern become?

In the table below, we show the calculation for
*v = 18 km/s, 24 km/s, 30 km/s and 600 km/s.*

| | | | | | |
|---|---|---|---|---|---|
| L | m | 10 | 10 | 10 | 10 |
| v | m/s | 18000 | 24000 | 30000 | 600000 |
| c | m/s | 300000000 | 300000000 | 300000000 | 300000000 |
| s | m | 1 | 1 | 1 | 1 |
| λ | m | 0,0000006 | 0,0000006 | 0,0000006 | 0,0000006 |
| | | | | | |
| a | m | 0,0006000 | 0,0008001 | 0,0010001 | 0,0200401 |
| b | m | 0,0006000 | 0,0007999 | 0,0009999 | 0,0199601 |
| | | | | | |
| a+b | m | 0,0012000 | 0,0016000 | 0,0020000 | 0,0400002 |
| a-b | m | 0,0000001 | 0,0000001 | 0,0000002 | 0,0000800 |
| | | | | | |
| d1=a+b | m | 0,0012000 | 0,0016000 | 0,0020000 | 0,0400002 |
| d2=2b | m | 0,0011999 | 0,0015999 | 0,0019998 | 0,0399202 |
| d3=a+b | m | 0,0012000 | 0,0016000 | 0,0020000 | 0,0400002 |
| d4=2a | m | 0,0012001 | 0,0016001 | 0,0020002 | 0,0400802 |
| | | | | | |
| y1=sλ/d1 | m | 0,0005000 | 0,0003750 | 0,0003000 | 0,0000150 |
| y2=sλ/d2 | m | 0,0005000 | 0,0003750 | 0,0003000 | 0,0000150 |
| y3=sλ/d3 | m | 0,0005000 | 0,0003750 | 0,0003000 | 0,0000150 |
| y4=sλ/d4 | m | 0,0005000 | 0,0003750 | 0,0003000 | 0,0000150 |

In the table above we show some calculation values.
*y1, y2, y3* and *y4* is the distance between two peaks in the interference pattern.

If you take *v = 30 km/s* as one did in the Michelson-Morley experiment, we see that the difference between the size of these distances is very small.

| | | | | | |
|---|---|---|---|---|---|
| y1-y2 | m | 0,00000003 | 0,00000003 | 0,00000003 | 0,00000003 |
| y2-y3 | m | 0,00000003 | 0,00000003 | 0,00000003 | 0,00000003 |
| y3-y4 | m | 0,00000003 | 0,00000003 | 0,00000003 | 0,00000003 |
| y4-y1 | m | 0,00000003 | 0,00000003 | 0,00000003 | 0,00000003 |

When turning the interferometer by 90° becomes $\Delta y = 0.000\,000\,030\,m$ which is *30 nanometers*. Could one read these small differences in 1887? Doubtful!

1 nm, a nanometer is equal to
0.000 000 001 meters and then above $\Delta y$ is
0.000 000 030 meters (30 nm)

It is no wonder that one could not determine how it is with the Earth's velocity towards the ether!
It is no wonder that one made the wrong conclusion about the ether! It's no wonder that it was a negative result!

What is most strange in my calculations is that for a value of $v$ you get approximately the same $\Delta y$ between the different sub-moments, when the interferometer is turned by 90°. And keep in mind that I have used $v = 30\,km/s$ and $600\,km/s$. It's 20 times more in the second case! This means that Michelson interferometer was **not suitable for use for its intended purpose**, to measure the Earth's velocity in space!

In this analysis, I have applied calculations that apply to 2-slot interference. In the Michelson-Morley experiment, one uses time difference between the two light rays. Below we also do this analysis.

Fig. 12-13 Exp. *1*   $\Delta t = (a-b)/c$
Fig. 14-15 Exp. *2*   $\Delta t = 0$
Fig. 16     Exp. *3*   $\Delta t = (a-b)/c$
Fig. 17-18 Exp. *4*   $\Delta t = 0$

This is calculated in a similar way as we calculated the spatial distance between $S_1$ and $S_2$.

We calculate $N = \Delta t / T$ where $T = 1,83*10^{-15}$ s.

Fig. 12-13 Exp. *1*   $N = 0,364 \ (0,40)$
Fig. 14-15 Exp. *2*   $N = 0$
Fig. 16     Exp. *3*   $N = 0,364 \ (0,40)$
Fig. 17-18 Exp. *4*   $N = 0$

In Exp. 2 and 4, there is no time difference between the two light rays, which means that no interference pattern should be created! Did it?

We look at the pictures in the above four experiments that for the light rays $S_1$ and $S_2$ there is both a spatial and temporal 'distance'. How are the interference patterns formed in these cases?

## Mathematics and SR

In physics, one uses the mathematics in a way that few other subjects do. For example, take the following formula:
　length = speed * time or
　$l = vt$

The unit of the physical quantities included in this formula is as follows:
- length (meters)
- speed (meter / second)
- time (second)

Take another example: calculation of area for a *right-angled* triangle.

　$A = ab/2$ where $a$, $b$ is the catheter of the triangle.

It is of utmost importance that both catheters have the same units.

I refer to *Introduction to Physics* by J.D. Curtnell and Kenneth W. Johnson, page 4:

"Only quantities with the same units can be added or subtracted."

Now we look in the book [14] pages 9-10: here a *right-angled* triangle is drawn where catheter A **represents the time** and has as unit *year*, the second catheter B **represents distance/length** and has as unit *light year*.

Furthermore, one calculates the hypotenuse, C, with something called "the modified Pythagas' theorem".

$C^2 = A^2 - B^2$

$A = 590/0{,}999$ year, $B = 590$ light year $\rightarrow C = 26$ year

Quote from [14]:
"If we allow us to modify Pythagoras' theorem of spruce, we can actually make it fit."
...
"What Einstein showed when he presented the special theory of relativity in 1905 was thus: Pythagas' theorem applies even in spacetime, though modified by a minus sign in front of the shortest catheter's square."

If I had done something like this in any of my math exams, I would never have received my diploma! No! You can't do that! One cannot change even "a little" in the existing rules, formulas, definitions, to get their results to match ...

## Physics and SR

See, if necessary, the chapter *Registration, calculation and transformation of coordinates* from page 20.

In the book *[3], page 125; Appendix;*
*A simple derivation of the Lorentz transformation*
they refer to the picture on page 66 (in that book).
"A light signal that runs along the positive x-axis is propagated according to the equation $x = ct$."
Similarly, you get $x' = ct'$. The light signal is propagated at the same speed $c$ also in the second coordinate system.

We show a similar picture below, Fig. 20.

Fig. 20

Here we have two reference systems S and S' that coincide at the beginning of the experiment.
On the x-axis an event occur at distance $x$ from both

reference systems.

The two reference systems S and S' will receive information about the event when the light signal reaches them.
We say that S and S' are stationary towards each other. Then we have the following situation, Fig. 21.

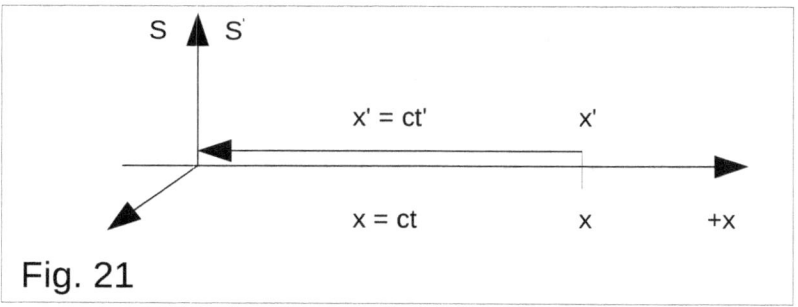

Fig. 21

In this case, we have $x = x'$ and $t = t'$.

Now we consider the following case: when S, S' are at the same point and their clocks are reset, an event occurs on the x-axis, S' starts to move to the right with the speed $v > 0$. The light signal will first meet the reference system S' and a little later S. In the moment when the light signal reaches S' the light signal passed the distance $x' = ct'$. And during the same time, S' have passed the distance $vt'$ to the right. The light signal continues toward S and pass the distance $x = ct$.

We illustrate this in Fig. 22.

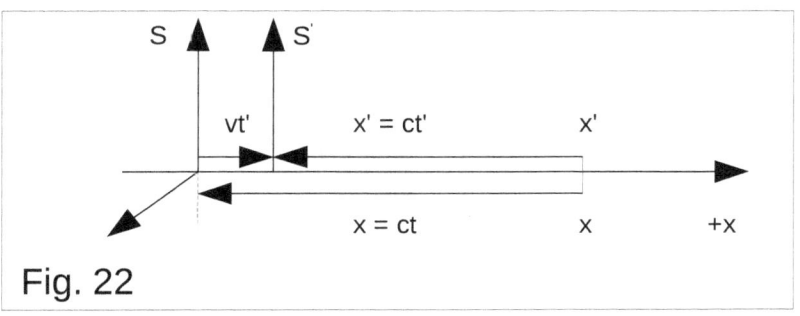

Fig. 22

From here we get
$$x = x'+vt'$$
or
$$ct = ct'+vt' \to ct = t'(c+v)$$
or
$$t = t'(c+v)/c \to t' = tc/(c+v)$$

These relationships between $x$ and $x'$ and also between $t$ and $t'$ result from a physical situation when two reference systems move towards each other at speed $v > 0$.

We get
$$x = x'+vt' \to x' = x-vt'$$
and
$$t' = tc/(c+v) \to x' = x-tcv/(c+v). \quad (1)$$

Than we have Lorentztransformationer:
$$x' = (x - vt)\gamma \qquad (LT_1)$$
$$t' = (t - vx/c^2)\gamma \qquad (LT_2)$$

We compare the first equation
$$x' = x\gamma - vt\gamma \qquad (LT_1)$$
with
$$x' = x - tcv/(c+v) \qquad (1)$$

This can be true only if $\gamma = 1$.

From here it results $v = 0$.


**This result, $v = 0$, is in contradiction to the theory's assumption that the two reference systems move toward each other at constant velocity $v > 0$!**

This shows that the special theory of relativity contains inaccuracies, is not self-consistent.

## Special theory of relativity and LT

We make a careful analysis of Lorentz transformations, LT, based on what they say and what the special theory of relativity says.

Lorentz transformations:

$$x' = (x - vt)\gamma \qquad (LT_1)$$
$$t' = (t - vx/c^2)\gamma \qquad (LT_2)$$

where $\gamma = 1/(1 - v^2/c^2)^{1/2}$ is called the Lorentz factor.

The special theory of relativity - SR - deals with two reference systems, S and S', which move at speed v > 0 against each other. It also says that nothing can move faster than light. Then we have the following conditions that must be followed when working with SR.

$$0 < v < c < +\infty$$

From here we get

$$0 < 1 < \gamma < +\infty$$

We consider an event that occurs in the two reference

systems. An observer in the reference system becomes aware of this event when the light signal from the event is recorded in the respective reference system. See Fig. 23.

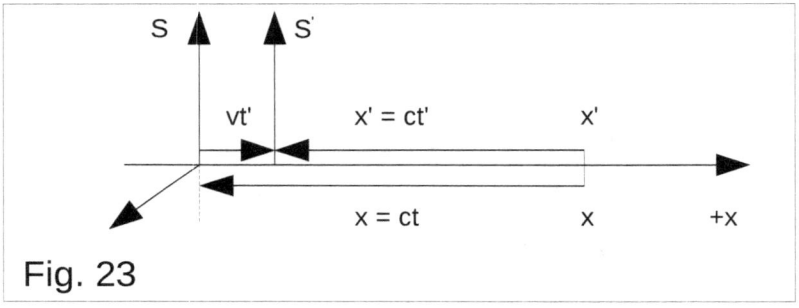

Fig. 23

Here we have additional conditions that must be followed when working with SR and Lorentz transformations – LT.

$x = ct,\ x' = ct'$

These are shown in the book *[3], page 125; Appendix; A simple derivation of the Lorentz transformation.*

Below we do an analysis of what happens if one of the variables $x,\ t,\ x',\ t'$ is set to zero.

a1) $x' = 0$
From $x' = ct'$ and $x' = 0 \to t' = 0$
From LT1, $x' = 0 \to 0 = (x - vt)\gamma \to x - vt = 0 \to x = vt$
From $x = vt$ and $x = ct \to ct = vt \to ct - vt = 0$
$\to t(c-v) = 0$
From $v < c \to c-v > 0 \to t = 0$
From $t = 0$ and $x = ct \to x = 0$

a2) $t' = 0$
From $x' = ct'$ and $t' = 0 \to x' = 0$
From LT1, $x' = 0 \to 0 = (x - vt)\gamma \to x - vt = 0 \to x = vt$
From $x = vt$ and $x = ct \to ct = vt \to ct - vt = 0$
$\to t(c-v) = 0$
From $v < c \to c-v > 0 \to t = 0$
From $t = 0$ and $x = ct \to x = 0$

a3) $x = 0$
From $x = ct$ and $x = 0 \to t = 0$
From LT1, $x = 0, t = 0 \to x' = (0 - 0)\gamma \to x' = 0$
From $x' = ct'$ and $x' = 0 \to t' = 0$

a4) $t = 0$
From $x = ct$ and $t = 0 \to x = 0$
From LT1, $t = 0, x = 0 \to x' = (0 - 0)\gamma \to x' = 0$
From $x' = ct'$ and $x' = 0 \to t' = 0$

These four results tell us that if one of the variables $x, t, x', t'$ is zero then everyone else must be zero! This also tells us that if one of the variables $x, t, x', t'$ is zero then both reference systems are **in a same point**. Then no Lorentz transformations are needed.

Above we only used Lorentz transformations - LT and conditions $x = ct$ and $x' = ct'$.

**Think about this for a while.**

What does the above result mean? This means that:

If there is an event $E = (x, t)$ in reference system S and if any of variables $x, t$ is zero then the second reference system S' will register an event $E' = (x', t') = (0, 0)$.

If there is an event $E' = (x', t')$ in reference system S' and if any of variables $x', t'$ is zero then the second reference system S will register an event $E = (x, t) = (0, 0)$.

We analyze a concrete example. See Fig. 24.

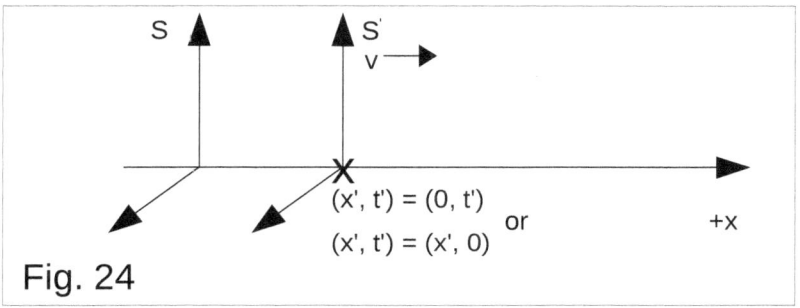

Fig. 24

If we have a situation as in Fig. 24, the result above means that Lorentz transformations - LT force the two reference systems to coincide in one.

See Fig. 25.

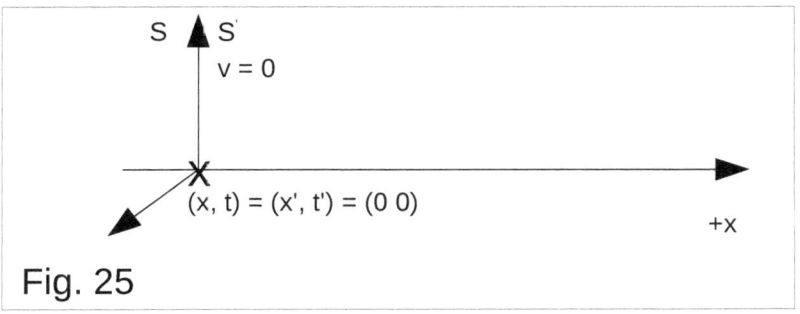

Fig. 25

We now analyze an example where all variables $x, t, x', t'$ are different from zero.

See Fig. 23. We have $x = x' + vt'$ or $ct = ct' + vt'$.

From $ct = ct' + vt' \to t = t'(c+v)/c$ and $t' = tc/(c+v)$
From LT1, $x' = ct'$, $x = ct$, $t' = tc/(c+v) \to ct' = (ct - vt)\gamma$
$\to c^2 t/(c+v) = t(c-v)\gamma$
We divide by $t$ and get
$c^2/(c+v) = (c-v)\gamma \to \gamma = c^2/(c^2-v^2) \to \mathbf{\mathit{v = 0}}$

Whatever we do, which individual cases we analyze, we come to the conclusion that $\mathbf{\mathit{v = 0}}$ ***or that Lorentz transformations apply only to the point*** $\mathbf{\mathit{(x, t) = (x', t') = (0, 0)}}$.

**We can say that Lorentz transformations have a built-in error or that their application area is only one point.**

They are the basis for the special theory of relativity and this means that this theory also has a built-in error.

The conclusion from this analysis tells us that the special theory of relativity is wrong in its entirety, **(is not self-consistent)**!

# Relativity with classical physics

We should now look at how we can calculate distance/ length between two points, in the two reference systems. See Fig. 26.

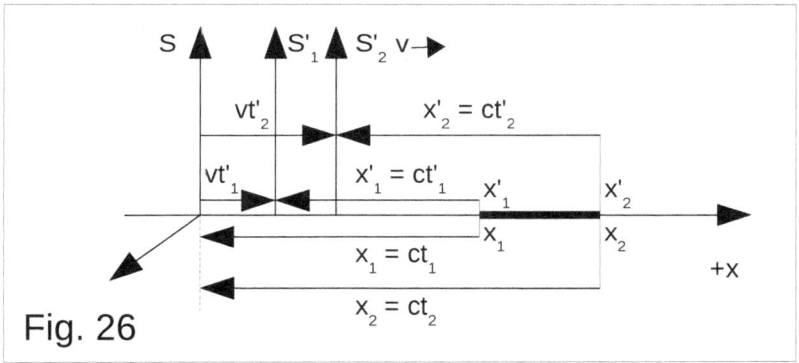

Fig. 26

We have two reference systems S and S'. S' moves to the right with speed $v > 0$. When the experiment begins, both reference systems are in the same point. Then, two light signals are transmitted from the points $x_1$ and $x_2$ on the x-axis. We can say that two events, $E_1$ and $E_2$, occur in the same time but in two different points on the x-axis.

For the reference system S we have:

$E_1 = (x_1, t_1)$ and $E_2 = (x_2, t_2)$.

The distance between $x_1$ and $x_2$ will then be

$$\Delta x = x_2 - x_1 = c(t_2 - t_1).$$

For the reference system S' we have:

$$E'_1 = (x'_1, t'_1) \text{ and } E'_2 = (x'_2, t'_2).$$

The distance between $x'_1$ and $x'_2$ will then be

$$\Delta x' = x'_2 - x'_1 = c(t'_2 - t'_1).$$

We'll calculate $\Delta x'$ using $\Delta x$.
From Fig. 26, it is clear that

$$x_1 = x'_1 + vt'_1 \rightarrow ct_1 = ct'_1 + vt'_1 = t'_1(c+v)$$
$$x_2 = x'_2 + vt'_2 \rightarrow ct_2 = ct'_2 + vt'_2 = t'_2(c+v)$$

$$ct_1 = t'_1(c+v) \rightarrow t'_1 = t_1 c / (c+v)$$
$$ct_2 = t'_2(c+v) \rightarrow t'_2 = t_2 c / (c+v)$$

Then we calculate $\Delta x' = x'_2 - x'_1 = c(t'_2 - t'_1) \rightarrow$
$$\Delta x' = c[t_2 c / (c+v) - t_1 c / (c+v)] \rightarrow$$
$$\Delta x' = c[(t_2 - t_1)(c / (c+v))] = \Delta x [c / (c+v)]$$

So we have
 $\Delta x' = \Delta x[c/(c+v)]$
 $\Delta x = \Delta x'[(c+v)/c]$.

We should now look at how we can calculate a time interval in the two reference systems. See Fig. 27.

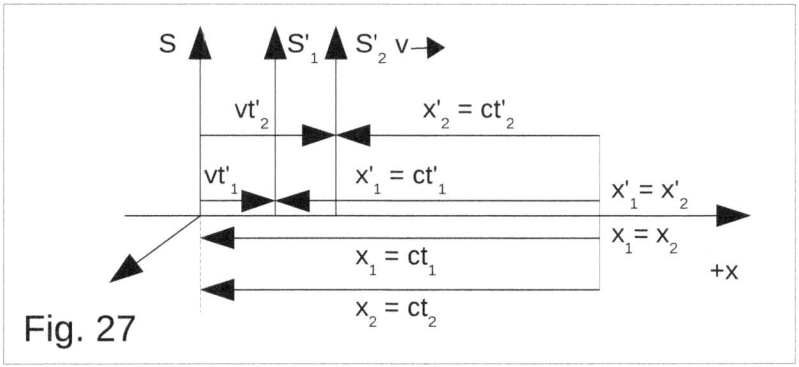

Fig. 27

We have two reference systems S and S'. S' moves to the right with speed $v > 0$. When the experiment begins, both reference systems are in the same point. Then, two light signals are transmitted from the point $x_1 = x_2$ on the x-axis. We can say that two events, $E_1$ and $E_2$, occur at the same point on the x-axis but with a time difference between them.

For the reference system S we have:
 $E_1 = (x_1, t_1)$ and $E_2 = (x_2, t_2)$.

The time difference between $t_1$ and $t_2$ will then be

$$\Delta t = t_2 - t_1.$$

For the reference system S' we have:

$$E'_1 = (x'_1, t'_1) \text{ and } E'_2 = (x'_2, t'_2).$$

The time difference between $t'_1$ and $t'_2$ will then be

$$\Delta t' = t'_2 - t'_1.$$

We'll calculate $\Delta t'$ using $\Delta t$.
From Fig. 27, it is clear that

$$ct_1 = t'_1(c+v) \rightarrow t'_1 = t_1 c/(c+v)$$
$$ct_2 = t'_2(c+v) \rightarrow t'_2 = t_2 c/(c+v)$$

Then we calculate $\Delta t' = t'_2 - t'_1 = (t_2 - t_1)[c/(c+v)] \rightarrow$
$\Delta t' = \Delta t[c/(c+v)]$

So we have

$$\boldsymbol{\Delta t' = \Delta t[c/(c+v)]}$$
$$\boldsymbol{\Delta t = \Delta t'[(c+v)/c)]}$$

We see that conversion factor:

   between S and S' is *(c+v)/c* and
   between S' and S is *c/(c+v)*.

But this does NOT mean that if the time interval measured in the two reference systems is different, the time varies in the ones.
It was NOT about any time dilation!

It was NOT that "time must go different quickly for observers moving in relation to each other".
This does NOT mean that if the length of a yardstick measured in the two reference systems is different, the yardstick has different lengths.
It was NOT about any length contraction!

And the light clock ticks the same way in a system at rest and in one that moves.

**Our impressions are relative but the reality is absolute!**

So conversion factor - we could call it the **relativity factor (rf)** - between S and S' and and vice versa is

$$rf_f\,(S',\,S) = c/(c+v)$$
$$rf_f\,(S,\,S') = (c+v)/c$$

But this only applies if the events we deal with appear "in front of S' ", or said otherwise if S' approaches the events we measure.
But if the events we deal with emerge "behind S' ", or said otherwise if S' moves away from the events we measure then relativity factor becomes another:

$$rf_b\,(S',\,S) = c/(c-v)$$
$$rf_b\,(S,\,S') = (c-v)/c$$

See chapter *Registration, calculation and transformation of coordinates* and picture Fig. 6.

Here we see that transformation of coordinates between two reference systems is not the same along the whole x-axis.

Transformation of coordinates between two reference systems moving at constant speed, $v > 0$, against each other **is not linear**!
In the various answers I received on my e-mail letters, some has stated that the special theory of relativity is used!

Therefore, I have wondered how it is possible that no anomaly is detected in all measurements.

In the table below we compare the value of

- Lorentz factor $\gamma = 1/(1 - v^2/c^2)^{1/2}$
- relativity factor $rf_f(S', S) = c/(c+v)$
- relativity factor $rf_f(S, S') = (c+v)/c$
- relativity factor $rf_b(S', S) = c/(c-v)$
- relativity factor $rf_b(S, S') = (c-v)/c$

| | |
|---|---:|
| c km/s | 300 000 |
| v km/s | 30 |
| Factor | Value |
| Lorentz factor $y = 1/(1-v^2/c^2)^{1/2}$ | 1,000000005 |
| relativity factor $rf_f(S', S) = c/(c+v)$ | 0,999900010 |
| relativity factor $rf_f(S, S') = (c+v)/c$ | 1,000100000 |
| relativity factor $rf_b(S', S) = c/(c-v)$ | 1,000100010 |
| relativity factor $rf_b(S, S') = (c-v)/c$ | 0,999900000 |

We see that when $v = 30 \text{ km/s}$ the difference between the Lorentz factor and the other factors is about $\pm 0.0001$.

## Derivation of LT

We follow the reasoning and the calculations from *[7], pages 14-15*. But this time we use our own notations.
See chapter Derivation of Lorentz transformations 1 in this book, page 34.

In the beginning you have two linear equations:

LE1: $x' = Ax+Bt$
LE2: $t' = Cx+Dt$

where A, B, C, D are constants.

To solve the above equation system, three special cases are used:

c1) The object in which event E occurs is in the S'-origin. $E' = (x', t') = (0, t')$
c2) The object in which event E occurs is in the S-origin. $E = (x, t) = (0, t)$
c3) One equates the object in which event E occurs with a light beam.

Of these, three pairs of conditions result. One replaces this conditions in LE1 and LE2. These three special cases then constitute the following equation systems:

SC1: $x' = 0$, $x = vt$
SC2: $x' = -vt'$, $x = 0$
SC3: $x' = ct'$ and $x = ct$

We replace these special cases in LE1 and LE2.

LE1, SC1: → $0 = Avt + Bt$
LE2, SC1: → $t' = Cvt + Dt$

LE1, SC2: → $-vt' = Bt$
LE2, SC2: → $t' = Dt$

LE1, SC3: → $ct' = Act + Bt$
LE2, SC3: → $t' = Cct + Dt$

Here we see that these results show different relationships between $t, t', A, B, C, D$.
You also get:

From LE1, SC1: → $0 = Avt + Bt$ → $B = -Av$
From LE1, SC2 and LE2, SC2 → $B = -Dv$
→ $D = A$

From LE1, SC3 and LE2, SC3 and $B = -Av$ and $D = A$
→ $C = -Av/c^2$

Finally, you get Lorentz transformations, LT.

LT1: $x' = (x - vt)\gamma$
LT2: $t' = (t - vx/c^2)\gamma$

where $\gamma = 1/(1 - v^2/c^2)^{1/2}$ is called the Lorentz factor.

I think everything didn't go as it should!
Therefore, we divide the derivation of LT into two parts:

Variant 1: $t = 0$
Variant 2: $t \mathrel{!}= 0$

Note that variant 1 is the case when both reference systems S and S' are in the same point and the time is zero and then we have

$(x', t') = (x, t) = (0, 0)$

Also note that in this case, $t = 0$,
our equations and special cases will look like this:

LE1: $x' = 0$
LE2: $t' = 0$

SCt: $t = 0$
SC1: $x' = 0, x = 0$
SC2: $x' = 0, x = 0$
SC3: $x' = 0, x = 0$

Then one must also realize that we can't do any derivation in this case!

Now we follow the derivation from [7] but with our own notations:
Variant 2: $t \mathrel{!}= 0$

That being said, we start over again:

LE1: $x' = Ax+Bt$
LE2: $t' = Cx+Dt$

where $A, B, C, D$ are constants.

SCt: $t \mathrel{!}= 0$
SC1: $x' = 0, x = vt$
SC2: $x' = -vt', x = 0$
SC3: $x' = ct', x = ct$

We replace special conditions, one by one, in the two linear equations and make the simplest calculations:

LE1, SC1: → $0 = Avt + Bt$
LE2, SC1: → $t' = Cvt + Dt$

LE1, SC2: → $-vt' = Bt$
LE2, SC2: → $t' = Dt$

LE1, SC3: → $ct' = Act + Bt$
LE2, SC3: → $t' = Cct + Dt$

We begin to analyze these results.

LE1, SC1: → $0 = Avt + Bt$ → $0 = t(Av + B)$

Now we **can** conclude that $Av + B = 0$, now this result is compelling! Because $t \mathrel{!}= 0$ **is obliged to** →

$Av + B = 0$ → $B = -Av$

**The result ResH1: $B = -vA$**

LE1, SC2: → $-vt' = Bt$
LE2, SC2: → $t' = Dt$

We replace $t' = Dt$ in $-vt' = Bt$ and get: $Bt = -Dvt$
Here we **can** divide by $t$ ($t \mathrel{!}= 0$): → $B = -Dv$

**The result ResH2: $B = -vD$**

From ResH1 and ResH2 → $D = A$

**The result ResH3: $D = A$**

Furthermore, we apply ResH1, ResH2 and ResH3 on

LE1, SC3: → $ct' = Act + Bt$
LE2, SC3: → $t' = Cct + Dt$

Then we get:

LE1, SC3: → $ct' = Act - Avt$
LE2, SC3: → $t' = Cct + At$

We multiply the whole equation $t' = Cct + At$ with $c$ and get

LE2, SC3: → $ct' = Cc^2t + Act$

From $ct' = Act - Avt$ and $ct' = Cc^2t + Act$ →

$Cc^2t = -Avt$, here we **can** divide by $t$ and sheep

$Cc^2 = -Av$ → $C = -vA/c^2$

**The result ResH4: $C = -vA/c^2$**

But note! Note! Note!
In this derivation in the book [7] one has not used equation

**LE2, SC1:** → $t' = Cvt + Dt$

You can't do so. You **must** use all equations and all conditions when solving a system of equations! Were one aware of this? Why has this equation not been used ?!

We replace $C = -vA/c^2$ and $D = A$ in the **forgotten equation**

$t' = Cvt + Dt.$

We make some calculations and get:

$t' = At(1 - v^2/c^2) \rightarrow t' = t/\gamma$

This result shows that time is **contracting** rather than **dilating**. SR says that $t' = t\gamma$, which is the formula for **time dilation, TD.** But if you replace the constants $C$ and $D$ in the forgotten equation and then replace $A = \gamma$, we get $t' = t/\gamma$.

How do relativists explain this result?

## Verification of LT

We follow the reasoning and the calculations from
*[7], page 14-15*. But this time we use our own notations.
See chapter Derivation of Lorentz transformations 1 in this book.

In the book [7], one starts from two linear equations

LE1: $x' = Ax+Bt$
LE2: $t' = Cx+Dt$

where $A$, $B$, $C$, $D$ are constants.

Three special cases are used:

SC1: $x' = 0$, $x = vt$
SC2: $x' = -vt'$, $x = 0$
SC3: $x' = ct'$, $x = ct$

You replace these special cases in LE1 and LE2 and get

LT1: $x' = (x - vt)\gamma$
LT2: $t' = (t - vx/c^2)\gamma$

where $\gamma = 1/(1 - v^2/c^2)^{1/2}$ is called the Lorentz factor.

But if Lorentz transformations are derived from LE1, LE2 using SC1, SC2, SC3 then these three special cases should verify LT1, LT2 without coming into contradiction.

We are now doing this verification. We replace special cases SC in Lorentz transformations LT.

LT1, SC1 → $0 = 0$
LT2, SC1 → $t' = t(1-v^2/c^2)\gamma \to t' = t/\gamma$

LT1, SC2 → $t' = t\gamma$
LT2, SC2 → $t' = t\gamma$

LT1, SC3 → $t' = t(1-v/c)\gamma$
LT2, SC3 → $t' = t(1-v/c)\gamma$

We see that this verification gives us **quite different** relationships between $t'$ and $t$!

According to my knowledge in mathematics, one should get a **mathematical equality** from each mathematical verification!
One wonders why one gets mathematical equality only in one of six verifications! It doesn't really feel great!

Below I give an example of similar cases we did in school:

We have a linear equation that represents a line in a coordinate system $(x, y)$.

LE1: $ax + by + c = 0$;

Say that this line intersects the y-axis in the point

$(x, y) = (0, 6)$

SC1: $x = 0; y = 6$

Say this line intersects the x-axis in the point

$(x, y) = (-4, 0)$

SC2: $x = -4; y = 0$

We derived our equation from LE1 using SC1 and SC2. You replace the value of $x$ and $y$ in LE1 and get the values for the constants $a, b, c$.

LT1: $3x - 2y + 12 = 0$

So we initially have a general equation LE1 and two special cases SC1, SC2 and from these we get the shape of our equation, our line.

LE1: $ax + by + c = 0$;
SC1: $x = 0; y = 6$
SC2: $x = -4; y = 0$
→
LT1: $3x - 2y + 12 = 0$

Now we want to be sure that we calculated correctly, so we verify our calculations.

Verification 1 (LT1, SC1): We replace SC1 i LT1.

$3(0) - 2(6) + 12 = 0 \rightarrow 0 - 12 + 12 = 0 \rightarrow \mathbf{0 = 0} \rightarrow OK$

Verification 2 (LT1, SC2): We replace SC2 i LT1.

$3(-4) - 2(0) + 12 = 0 \rightarrow -12 - 0 + 12 = 0 \rightarrow \mathbf{0 = 0} \rightarrow OK$

In both verifications we got as a result a **mathematical equality**!

So it should be even when solving a system of two equations or more. So it should be even when verifying the derivation of the LT!

We show our school example in Fig. 28.

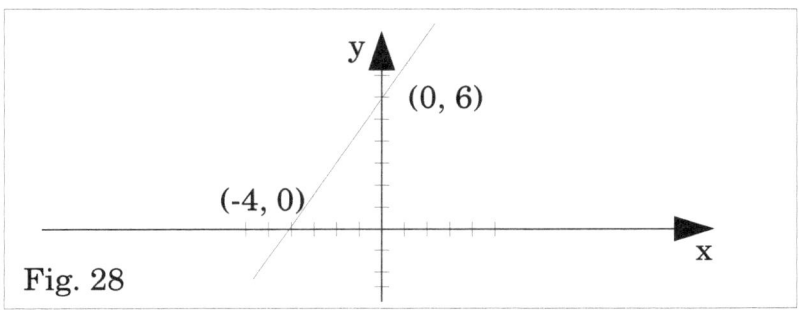

Fig. 28

Now we return to the verification of the derivation of LT.

We've got the following results:

ResV1: LT1, SC1 → $0 = 0$
ResV2: LT2, SC1 → $t' = t/\gamma$
ResV3: LT1, SC2 → $t' = t\gamma$
ResV4: LT2, SC2 → $t' = t\gamma$
ResV5: LT1, SC3 → $t' = t(1-v/c)\gamma$
ResV6: LT2, SC3 → $t' = t(1-v/c)\gamma$

How is this possible? Why do we get different relationships between $t'$ and $t$? Why don't we get a **mathematical equality** for each verification?

The verification of LT gives different formulas for the

relation between $t'$ and $t$ depending on how the experiment is arranged.

In the literature, check the term time dilation, TD. The special theory of relativity says that the time in the reference system that moves is

TD: $t' = t\gamma$

We see that this only applies in 1 of 3 special cases, or in 2 out of 6 verification results!

How to interpret this? That the time dilation, TD, and with it even length contraction, LC, is dependent on where the events we analyze occur?

I've read a lot about the special theory of relativity but never have I encountered such an explanation!
In any case, they don't boast about it!

SR = The special theory of relativity!
Above verification can confirm that it is "true"!
SR is **very special**, it only applies in 1 of 3 special cases!

These are the special cases with which LT was derived. And LT is the foundation of the whole building of SR!

I am actually quite surprised at this!
How is it possible that mathematicians and physicists have allowed this to continue throughout these years? I do not want to say that I am an expert in mathematics. I am a "regular" mathematician. To analyze SR, no extraordinary knowledge of mathematics is needed!

But in all of the above parts of SR that I have analyzed, I come to the conclusion that LT applies only to a single point, or that v = 0, or to other results that are more or less strange or unexplained!

Could it be that I am wrong in **all** my calculations, conclusions? It is unlikely! Because if any scientist / relativist states that I am wrong, you have to show that I am wrong everywhere. Because it is enough that I am right in one part of my analysis, then the whole theory of relativity falls!


In conclusion, I would like to show you a logical analysis of the derivation of LT.


We have two linear equations:

LE1: $x' = Ax + Bt$
LE2: $t' = Cx + Dt$

We have three special cases:

SC1: $x' = 0$, $x = vt$
SC2: $x' = -vt'$, $x = 0$
SC3: $x' = ct'$, $x = ct$

This results in LT:

LT1: $x' = (x - vt)\gamma$
LT2: $t' = (t - vx/c^2)\gamma$

Consider these parts of the derivation! Check the previous chapter Derivation of LT.

LE1, SC1:
→ The result ResH1: $B = -vA$

LE1, SC2 and LE2, SC2:
→ The result ResH2: $B = -vD$

From ResH1 and ResH2
→ The result ResH3: $D = A$

Furthermore, one appliesResH1, ResH2 and ResH3 on LE1, SC3 and LE2, SC3

→ The result ResH4: $C = -vA/c^2$

And finally you get LT:

LT1: $x' = (x - vt)\gamma$
LT2: $t' = (t - vx/c^2)\gamma$

Now consider the three special cases (again and again):

SC1: $x' = 0, x = vt$
SC2: $x' = -vt', x = 0$
SC3: $x' = ct', x = ct$

Mathematically, these special cases constitute three equation systems on two equations each. But if you look closely you see that these three equation systems are **incompatible** with each other!

SC1, SC3: → $v = c$
SC2, SC3: → $v = -c$
SC1, SC2: → $x' = 0, t' = 0, x = 0, t = 0$

That is why my analysis always came to the conclusion that LT did **not match** reality!
SR is based on LT which only applies in the point of the beginning of the experiment, when both reference

systems are in the same point $(x', t') = (x, t) = (0, 0)$.

And when the two reference systems are in the same point then it is pointless to talk about any transformations between coordinates between these two reference systems!
We look again at the three special cases

   SC1, SC3: → $v = c$
   SC2, SC3: → $v = -c$
   SC1, SC2: → $x' = 0, t' = 0, x = 0, t = 0$

What do you think?
Because I can't believe in the truth of SR after all this analysis with its results!
Herewith I urge physics and mathematics researchers to review my calculations, my ideas, my results!

Herewith I urge relativists to defend their claims!
And show that I'm wrong!
Or admit that I'm right!
Because

   $0 = 0$ is true
   $1 = 0$ is false

and everything in between is also false!

## SR and spacetime

In the special theory of relativity, the space *(x, y, z)* has been merged with time *(t)* in a new concept **spacetime**. But there is NO physical connection between *(x, y, z)* and *(t)*!
Then a point in spacetime is denoted by *(x, y, z, t)*.

We are used to representing different objects in a plane, on paper, for example. Then you can represent lines, figures that lie in a plane.

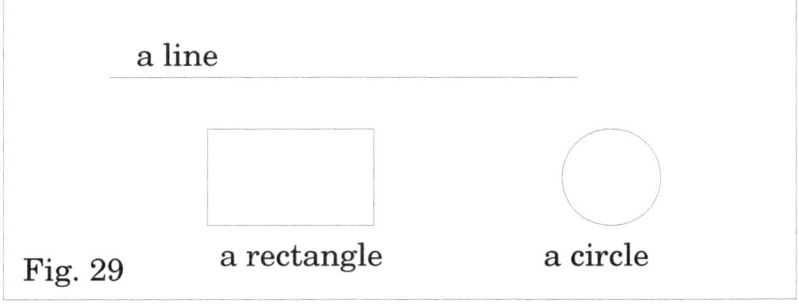

Fig. 29

It is more difficult to work with three-dimensional figures.
Imagine a cube, a pyramid, a globe.

And it is **impossible** to represent the spacetime on paper! We can't draw all 4 dimensions!

But individual parts of the spacetime can be drawn on the paper.

Imagine an experiment where the studied object moves only on one line, in a plane, such as the x-axis.

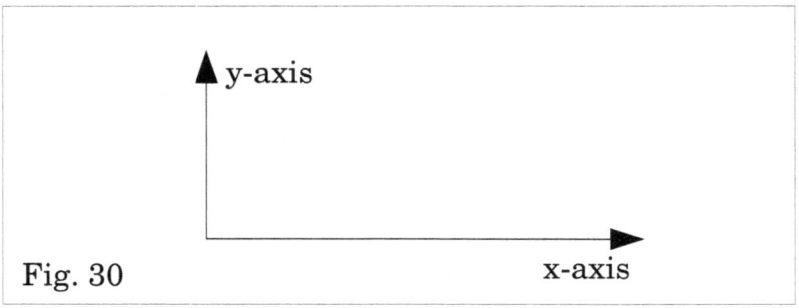

Fig. 30

In Fig. 30, we have a 2-dimensional coordinate system.

Now we are turning this plan into a spatial appearance.

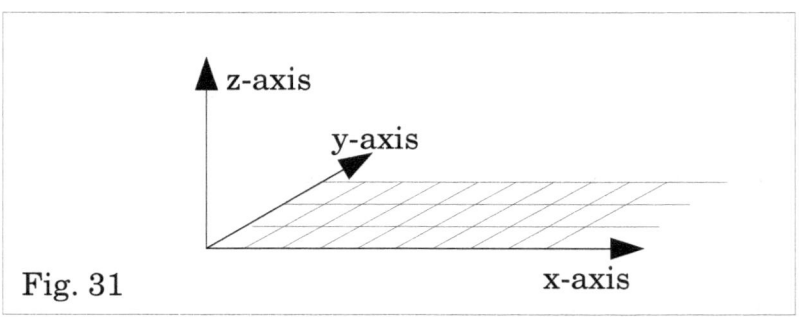

Fig. 31

Now we remove the z-axis and replace it with the time coordinate, t-axis. So, if we imagine two events taking place in space time at time $t = 0$ and $t = t'$, they will be displayed on our image as in Fig. 32.

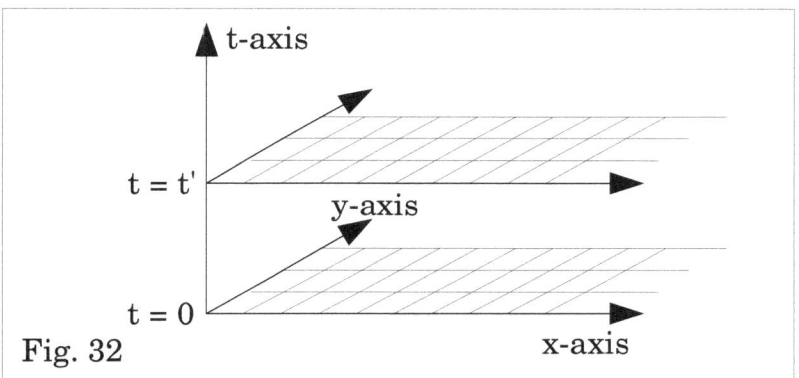

Fig. 32

**Note: But here one has made a mathematical model that does not match reality!**

We now depict an object that moves between two points on the x-axis, or a parallel line with the x-axis. We mark these two points with J (Earth) and B (star Betelgeuse).
See Bibliography *[14]*.

See Fig. 33. Imagine a space ship starting in J at time $t = 0$ and reaching B at time $t = t'$.

When we draw this, it will look like in Fig. 33. When the spaceship starts at point J, it is in the "lower" plane and when the spaceship arrives at point B it is in the "upper" plane.

But in reality it is the same plan! It's just our model that looks like that. It is only our representation of the 4 dimensions that confuse us.

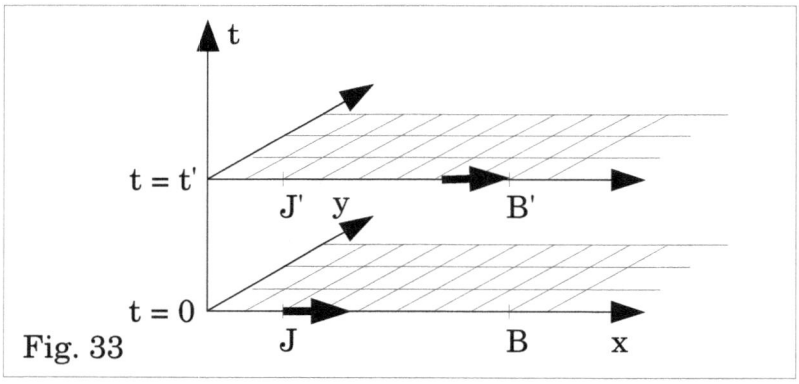

Fig. 33

In SR, one says that the spaceship has moved in spacetime as shown in Fig. 34.

SR says the spaceship moves in spacetime along the JB' section. In that model. But not in reality! The worst thing is that you equate the time axis, $t$, with a length axis, $x$. And that can be done in the model, but what does it have to do with reality?

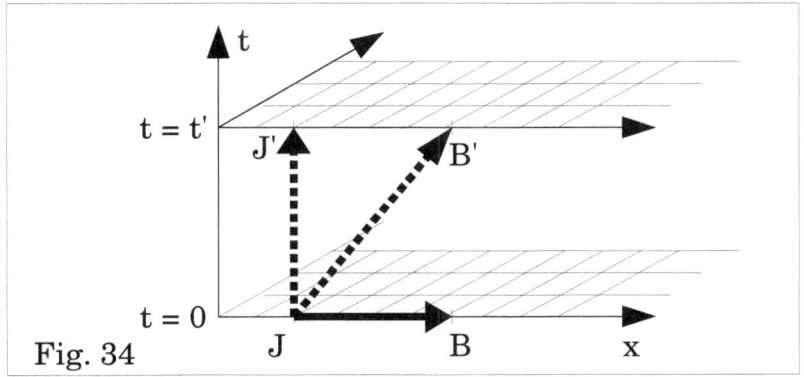

Fig. 34

The spacetime that is described in SR is an abstract concept and has nothing to do with our reality!

If you make a mathematical model of reality then you should ensure that the model corresponds to reality!

You can draw what we did in Fig. 33 and Fig. 34, but you cannot make any calculations whatsoever.

Looking at the book [14] page 9, a similar triangle is shown as the one we have in Fig. 34, the triangle JBB'.

The problem is that in that book one calculates the diagonal (hypotenuse) using something called the modified Pythagas' theorem:

$$(\text{Hypotenuse})^2 =$$
$$(\text{Longest catheter})^2 - (\text{Shortest catheter})^2$$

The longest catheter has as unit *year*, unit of time!
The shortest catheter has as unit *light year*, unit of length!

I refer to *Introduction to Physics* by J.D. Curtnell and Kenneth W. Johnson, page 4:
"Only quantities with the same units can be added or subtracted."

And then you change the Pythagas' theorem!
With what right do you do it?

Two huge errors!

You can't mix apples and pears in a mixer and think you get pure orange juice! It's not possible!

# Lorentz factor and its value in different points of spacetime

We consider Lorentz transformations, LT, below.

LT1: $x' = (x - vt)\gamma$
LT2: $t' = (t - vx/c^2)\gamma$

where $\gamma = 1/(1 - v^2/c^2)^{1/2}$ is called the Lorentz factor.

The Lorentz factor is a function of speed $v$.
If $v = 0$ then we have $\gamma = 1$ otherwise we have $\gamma > 1$.
$\gamma$ is always $\neq 0$.

In *[7], page 14-15* they derive above LT and as a condition they have **$v > 0$**.

LT is derived from two linear general transformations/equations

LE1: $x' = Ax+Bt$
LE2: $t' = Cx+Dt$.

Three special cases are used for this derivation:
SC1: $x' = 0, x = vt$
SC2: $x' = -vt', x = 0$
SC3: $x' = ct', x = ct$

We can rewrite Lorentz transformations.
G for gamma; LT for Lorentz transformations.

GLT1: $\gamma = x'/(x - vt)$, $x - vt \neq 0$
GLT2: $\gamma = t'/(t - vx/c^2)$, $t - vx/c^2 \neq 0$

We can calculate the value of $\gamma$ using the variables $x'$, $t'$, $x$, $t$, $v$ and the constant $c$.

We calculate the value of $\gamma$ using just the three special cases used in the derivation of Lorentz transformations.

We now make this calculation by replacing special cases SC in Lorentz transformations, LT.
From the points that verify SC1-SC3, we choose the ones that have $t > 0$, $t' > 0$ or $t \neq 0$, $t' \neq 0$.

If we think, it is not possible to derive LT when $t = 0$ or $t' = 0$.

Because then SC1, SC2 and SC3: $x' = 0$, $x = 0$.
So conditions $t > 0$ and $t' > 0$ should exist as initial conditions in the derivation of LT.

GLT1, SC1: $\gamma = x'/(x - vt) = 0/(vt - vt) = 0/0$
(mathematical nonsense)

GLT2, SC1: $\gamma = t'/(t - vx/c^2) = t'/(t - vvt/c^2) =$
$= (t'/t)(1/(1- v^2/c^2)) = \gamma^2(t'/t) \to$
$\gamma = \gamma^2(t'/t) \to \gamma = t/t'$

GLT1, SC2: $\gamma = x'/(x - vt) = -vt'/(0 - vt) =$
$= -vt'/-vt = t'/t \to \gamma = t'/t$

GLT2, SC2: $\gamma = t'/(t - vx/c^2) = t'/(t - 0) = t'/t$
$\to \gamma = t'/t$

GLT1, SC3: $\gamma = x'/(x - vt) = ct'/(ct-vt) =$
$= ct'/(t(c-v)) \to \gamma = t'c/t(c-v)$

GLT2, SC3: $\gamma = t'/(t - vx/c^2) = t'/(t-vct/c^2) =$
$= t'/t(1-v/c)$
$\to \gamma = t'c/t(c-v)$

We summarize the results of these calculations:

GLT1, SC1: $\gamma = 0/0$ (mathematical nonsense)
GLT2, SC1: $\gamma = t/t'$

GLT1, SC2: $\gamma = t'/t$
GLT2, SC2: $\gamma = t'/t$

GLT1, SC3: $\gamma = t'c/t(c-v)$
GLT2, SC3: $\gamma = t'c/t(c-v)$

We see that the Lorentz factor gets different mathematical expressions for different points of spacetime.

The above results give us four different expressions for Lorentz factor:

R1: $\gamma = 0/0$ (mathematical nonsense)
R2: $\gamma = t/t'$
R3: $\gamma = t'/t$
R4: $\gamma = t'c/t(c-v)$

But the Lorentz factor is the function of **only** $v$. So whatever point of the spacetime we use, we must get the same value!

We exclude R1: $\gamma = 0/0$ (mathematical nonsense), but already this result tells us something is wrong!
Why should we not be able to calculate the value of $\gamma$ for a point in spacetime used to derive Lorentz transformations?
LT applies to all points in spacetime. So we should be able to calculate the value of $\gamma$ at any point. And we should not come to mathematical nonsense!
The mathematics says that one cannot translate a physical phenomenon into a mathematical model anyway and then draw all possible conclusions from

this model.
**Mathematics is the queen of science!**

Analysis of other cases gives the following three variants:

V1)
    R2: $\gamma = t/t'$
    R3: $\gamma = t'/t$
    $\rightarrow t/t' = t'/t \rightarrow t = t' \rightarrow \gamma = 1 \rightarrow v = 0$

V2)
    R3: $\gamma = t'/t$
    R4: $\gamma = t'c/t(c-v)$
    $\rightarrow t'/t = t'c/t(c-v) \rightarrow c/(c-v) = 1 \rightarrow v = 0$

V3)
    R2: $\gamma = t/t'$
    R4: $\gamma = t'c/t(c-v)$
    $\rightarrow t/t' = t'c/t(c-v) \rightarrow (t/t')^2 = c/(c-v)$
    $\rightarrow \gamma^2 = c/(c-v) \rightarrow c^2/(c^2-v^2) = c/(c-v)$
    $\rightarrow c/(c+v) = 1 \rightarrow v = 0$

All these calculations of the value for Lorentz factor show that we either come to mathematical nonsense or that $v = 0$.
From here we conclude that the derivation of Lorentz transformations is incorrect (is not self-consistent).

## Twin Paradox:
## ... and so they lived ... at the same age ... ever after

Quote:
*"At the same time, these are the worlds we have the hardest to understand, worlds where illustrative models deceive us and we find paradoxes. But there is only one world and it has no paradoxes. Only our models that hold paradoxes "*
*[Vid skiljevägen : essäer om människan och hennes framtid; Ulf Sinnerstad; 2006, swedish]*

My motto:
*When we study physical phenomena, we always make a mathematical model of them. In such a model, there are built-in physical laws that are held together by mathematical tools. If the description of the physical phenomenon is correct, the mathematical model is also correct!*

How is it then? Are there any paradoxes in our world? Or are there any paradoxes only in the models one create to try to explain how our world works?

We consider the following thought experiment:

## Partial experiment 1

Two observers of the same age, O and O', are in the same place. They agree that O' gets on a trip and returns back. They wonder how it will be with their age when they meet again.

We make a **mathematical model** of this thought experiment.
O and O' agree that their relative velocity should be $v > 0$ and that they communicate with light signals.
O' moves at constant speed $v$ in a straight line.

The two observers constitute two reference systems, S and S'. When we draw this, S' will move to the right on the x-axis. In order to follow the evolution of time, one will send a light signal from both reference systems every second. We share the x-axis in equal stretches of the length $vt$, where $t$ is the unit of time.
For example: if the speed is $v = 30\ km/s$ this distance will be $30\ km$.

The experiment is designed such that each time S' pass a distance $v$, it sends a light signal. This corresponds to the deal that S' sends a light signal each unit time (in our case every second).

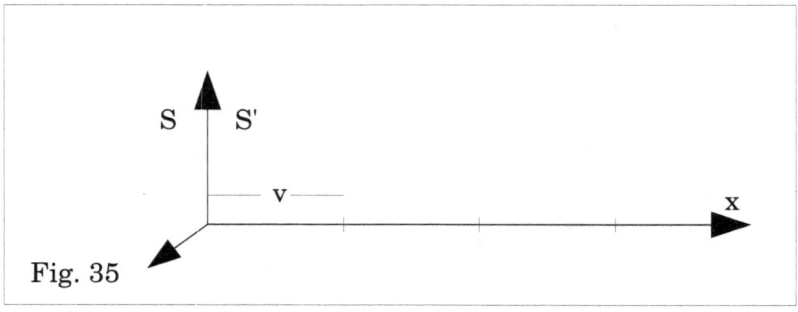

Fig. 35

S will record the time when the light signal from S' comes to S.

In Fig. 36 we see what happens when the first light signal is transmitted.

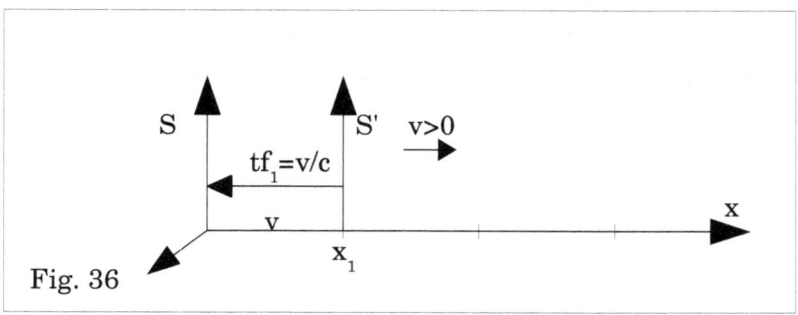

Fig. 36

We now calculate the time when this first light signal reaches S. We can intuitively see that the time $kf_1$ becomes ($kf$ = time forward, when S' moves forward):

$kf_1 = 1 + v/c$

How is this explained? The time $kf_1$ is equal to *1 second* (the time during which S' passed distance $v$) plus the time the light signal needs to go the same distance.
The light signal moves at the speed of light $c$, which approximates to *300,000 km/s*.

In Fig. 37 we do the same for step 2, when the light signal is transmitted from point $x_2$ at distance $2v$ from S-origo.

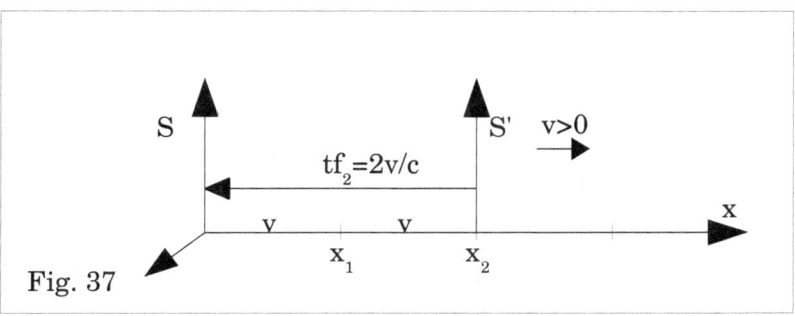

Fig. 37

The time when the second light signal reaches S becomes:

$kf_2 = 2 + 2v/c$

Similarly, we can write the formula for the step $i$:

$$kf_i = i + iv/c$$

We say that S' moves away from S in 10 steps. So when S' reaches its goal, the distance to S becomes $10v$.

$$kf_{10} = 10 + 10v/c$$

How does S perceive this?
S knows that the first signal is transmitted after the first second S' left S.
But this signal is a bit late. The delay time is $v/c$ seconds.
The second signal comes with twice as much delay, $2v/c$ seconds.

The light signals will be sparse and sparse. But that's nothing strange.
One can compare this with the phenomenon of the **red shift of light** when a light source go away.
Here we enter a more compact formula for the time S records:

$$kf_i = i + iv/c \rightarrow kf_i = i(1 + v/c) \rightarrow kf_i = i(c + v)/c \rightarrow$$
$$\boldsymbol{kf_i = i(c + v)/c}$$

## Partial experiment 2

Now we look at what happens when S' returns and begins approaching S at constant speed $v$. When S' passed the first distance $v$ on the way back, a light signal is transmitted. See Fig. 38.

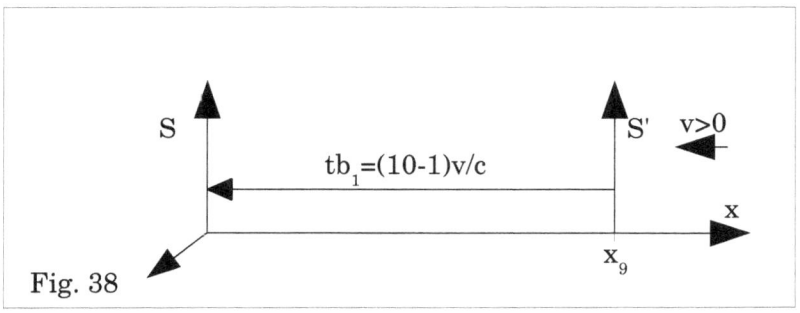

Fig. 38

We denote the time that S registers with $kb$ ($kb$ = the time backwards, when S' goes backwards, back to S).

$kb_1 = 10 + 1 + tb_1 \rightarrow kb_1 = 10 + 1 + (10-1)v/c$

Now we show the moment when S' transmits the penultimate light signal. Then S' is at a distance $v$ from S.

$kb_9 = 10 + 9 + tb_9 \rightarrow kb_9 = 10 + 9 + (10-9)v/c$

- 123 -

The general formula for the time when S' returns will be:

$kb_i = 10 + i + tb_i \rightarrow kb_i = 10 + i + (10-i)v/c \rightarrow$
$kb_i = 10 + 10v/c + i - iv/c \rightarrow kb_i =$
$\quad = 10(c+v)/c + i(c-v)/c \rightarrow$
$\boldsymbol{kb_i = 10(c+v)/c + i(c-v)/c}$

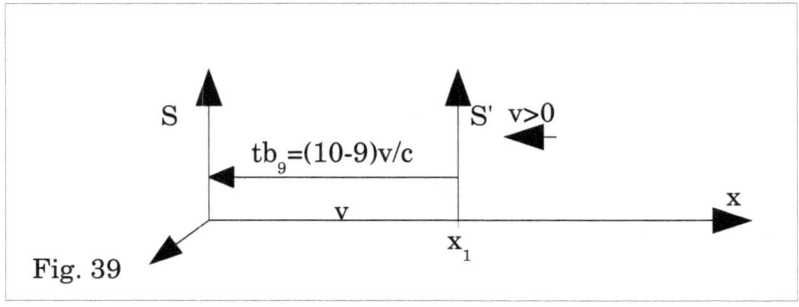

Fig. 39

How does S perceive this?
We look intuitively that when S' approaches S, light signals will be recorded in S at a faster rate. One can compare this with the phenomenon of the **blue shift of light**, when a light source approaches the observer.

The last light signal will be

$kb_{10} = 10 + 10 + tb_{10} \rightarrow kb_{10} = 10 + 10 + (10-10)v/c$
$\rightarrow$

$kb_{10} = 10 + 10 \rightarrow kb_{10} = 20$

This is quite normal. S' needed 10 seconds to reach his goal and 10 seconds to return.

In the case of S, observer O will not notice anything strange.
It is only the time interval between which O registers the light signals from S':
bigger and bigger at S' journey towards the goal and shorter and shorter at S' return.

S will not say that time goes faster or slower.
The time in S will tick at the same rate.

## Partial experiment 3

How is it then for S'? We initially said that even S sends a light signal every second.
S' will record these signals.

We show in Fig. 40 the moment when S sends the first signal. The first light signal is transmitted when S' is at distance $v$ from S.
Note that while the light signal moves towards S' hence this moves a bit to from point $x_1$.

We now calculate the time when this first light signal

from S reaches S'. We can intuitively see that the time $kf'_1$ becomes ($kf'$ = the time forwards, when S' goes forward):

$$kf'_1 = 1 + tf'_1 \rightarrow kf'_1 = 1 + v/(c-v)$$

How is this explained? The time $kf'_1$ is equal to *1 second* (the time during which S' passed distance $v$) plus the time the light signal needs to go the same distance, plus extra time to catch the S'.

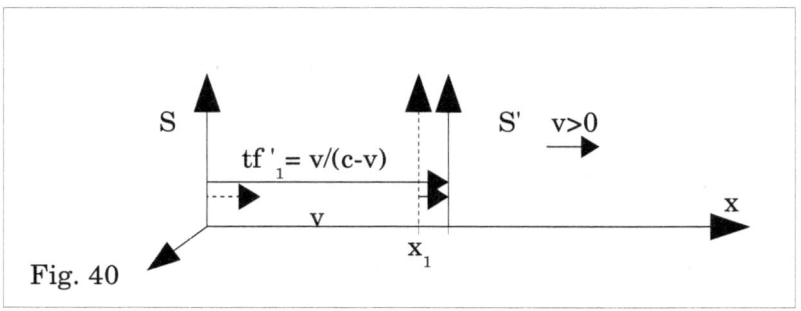

Fig. 40

$$ct = v + vt \rightarrow ct - vt = v \rightarrow t(c - v) = v \rightarrow t = v/(c - v)$$

In Fig. 41 we do the same for step 2. The light signal is transmitted from S at the moment S' is at a distance of $2v$ from the S-origo.

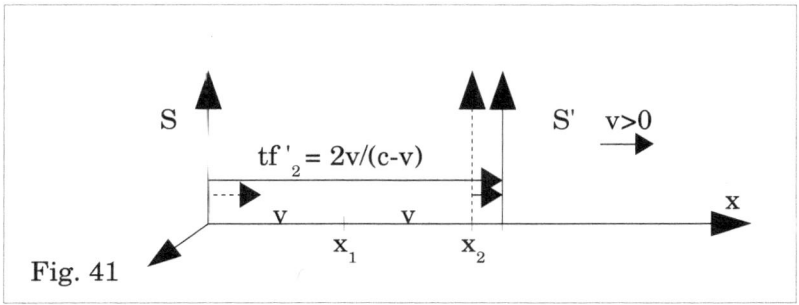

Fig. 41

The time when the second light signal reaches S' becomes:

$$kf'_2 = 2 + tf'_2 \rightarrow kf'_2 = 2 + 2v/(c-v)$$

Similarly, we can write the formula for the step $i$:

$$kf'_i = i + iv/(c-v)$$

S' moves away from S in 10 steps. So when S' reaches its goal, the distance to S becomes $10v$.

$$kf'_{10} = 10 + 10v/(c-v)$$

How does S' perceive this?

S' know that the first signal is transmitted after the first second S' left S.
But this signal is a bit late. The delay time is $v/(c-v)$

seconds.
The second signal comes with twice as much delay, $2v/(c-v)$ seconds.

The light signals will be sparse and sparse. But that's not strange.
One can compare this with the phenomenon of the **red shift of light** when a light source go away from us . Here we enter a more compact formula for the time S' records:

$$kf'_i = i + iv/(c-v) \rightarrow kf'_i = i(1 + v/(c-v)) \rightarrow kf'_i = ic/(c-v) \rightarrow$$
$$\boldsymbol{kf'_i = ic/(c-v)}$$

S' perceives the same phenomenon as S perceived when the two reference systems went from each other.

## Partial experiment 4

Now we look at what happens when S' returns and begins to approach S at constant speed $v$. When S' passed the first distance $v$ on the way back, a light signal is transmitted from S. See Fig. 42.

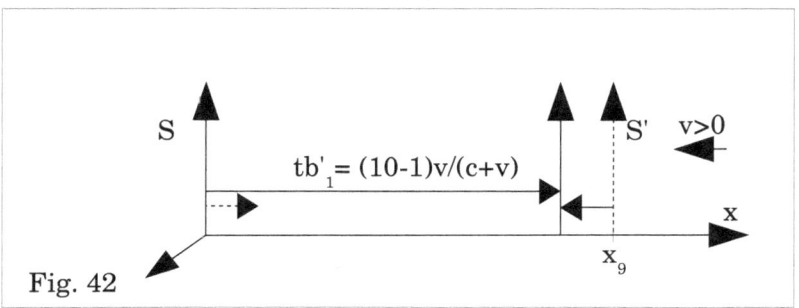

Fig. 42

We denote the time that S' registers with $kb'$ ($kb'$ = the time backwards, when S' goes backwards, back to S). First, we will explain how we came to the term for $tb'_1$. When S' is in $x_9$, when it has passed distance $v$ from the final target, a light signal is transmitted from S. At this point, the two reference systems are at a distance of $9v$. Meanwhile the light signal approaches S', this reference system moves a bit. So we have the following relationship:

$$9v = ct + vt \to t(c+v) = 9v \to t = 9v/(c+v) \to t = (10-1)v/(c+v)$$

Then the time when S' records the first light signal after it turns back is

$$kb'_1 = 10 + 1 + tb'_1 \to kb'_1 = 10 + 1 + (10-1)v/(c+v)$$

Now we show the moment when S sends the penultimate light signal. Then S' is at a distance $v$ from S. See Fig. 43.

The time when S' registers the penultimate light signal after it turns back is

$$kb'_9 = 10 + 9 + tb'_9 \rightarrow kb'_9 = 10 + 9 + (10-9)v/(c+v)$$

The general formula of the time recorded by S' at the time of return will be:

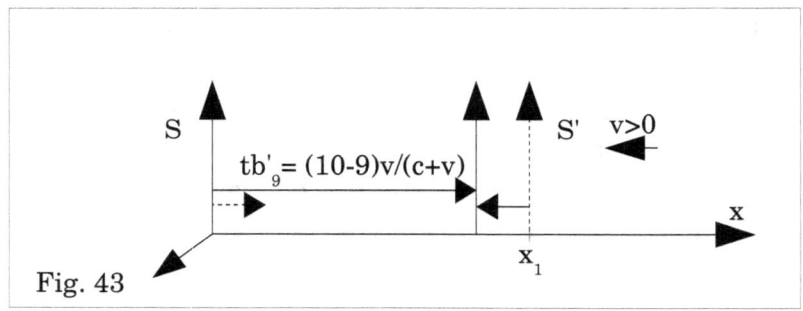

Fig. 43

$$kb'_i = 10 + i + tb'_i \rightarrow kb'_i = 10 + i + (10-i)v/(c+v) \rightarrow$$
$$kb'_i = 10 + 10v/(c+v) + i - iv/(c+v) \rightarrow$$
$$\boldsymbol{kb'_i = 10 + 10v/(c+v) + ic/(c+v)}$$

How does S' perceive this?

We see intuitively that when S' approaches S, light signals will be recorded in S' at a faster rate. One can compare this with the phenomenon of the **blue shift of light** when a light source approaches the observer.

The last light signal will be

$kb'_{10} = 10 + 10 + tb'_{10} \rightarrow kb'_{10} = 10 + 10 + (10-10)v/(c+v) \rightarrow$
$kb'_{10} = 10 + 10 \rightarrow kb'_{10} = 20$

This is quite normal. S' needed 10 seconds to reach his goal and 10 seconds to return.

In the case of S', the observer O' will not notice anything strange.
Only the time interval between which O' records the light signals from S.
Bigger and bigger at S's journey towards the goal and shorter and shorter at S's return.

O' will not perceive that time goes faster or slower. The clock in O' will tick at the same rate.

## Summary

We will summarize these partial experiments by writing the general formulas for
time of registration of light signals that reach both S and S'.

$kf_i = i(c + v)/c$
$kb_i = 10(c+v)/c + i(c-v)/c$
$kf'_i = ic/(c-v)$
$kb'_i = 10 + 10v/(c+v) + ic/(c+v)$

We look specifically at the term containing $i$ (where $i$ are step number for part experiments.

$i(c + v)/c$
$i(c - v)/c$
$ic/(c - v)$
$ic/(c + v)$

Not a single part experiment is similar to the other based on these expressions.
This I just point out as a curiosity.

Similar expressions for $c$ and $v$ were obtained on pages 23-24 and 88.

## Conclusion:

When the twin O' returns home to O, the clocks show the same time and they match the expected time of the return of the twin.

Note that both O and O' have their own clocks ticking at the same rate.
The reader and especially an adept of the special theory of relativity will argue against this statement.

How can we ensure that the clocks in both reference systems tick at the same rate?

Imagine how the above thought experiments are designed:
In S we have a clock ticking every second. Here's nothing to discuss, because S is the stationary reference system.
We mentioned that S' also sends a light signal every second. How can we be sure it's the same "second" as in S, that it's the same time?
We have divided the distance at which S' moves in equal distances.
If S' moves at velocity $v\ km/s$, then the length of these "distance unit" is equal to $v\ km$. Exactly when S' passes this milestone, his clock ticks.

In addition, each observer measures the time when the light signal from the twin arrives. These are the times we have calculated in the above four parts experiments.
That are these times that are interpreted as that the time of the others is slower.

**But it's completely wrong to think so.**

When twins are going away from each other, both will perceive the same thing, that the recorded signals from the other become sparse and sparse as the time passes.
When S' turns and the two approach each other, the recorded signals from the other will "tick" faster and faster and the last "tick" will coincide exactly with "tick" from the real clock.

## The twin paradox: The third brother

We continue the analysis of the twin paradox that we began in chapter Mathematics and SR on page 71.

There we have seen how the author of the book [14] presents the problematics of the time dilatation and the twin paradox. We have seen how they use "the modified Pythagoras' theorem" to calculate the hypotenuse in a right-angled triangle.

$C^2 = A^2 - B^2$

$A = 590/0{,}999$ *year*, $B = 590$ *light year* $\rightarrow C = 26$ *year*

You see in Fig. 44 the right-angled triangle ABC.

E = Planet Earth
B = Star Betelgeuse

Brother $B_0$ stays on Earth while brother $B_2$ flies at a rate of 99.9% of the speed of light to star Betelgeuse and back to Earth.

We now look at what happens if the third brother flies to planet X which is half the distance between Earth and Betelgeuse.

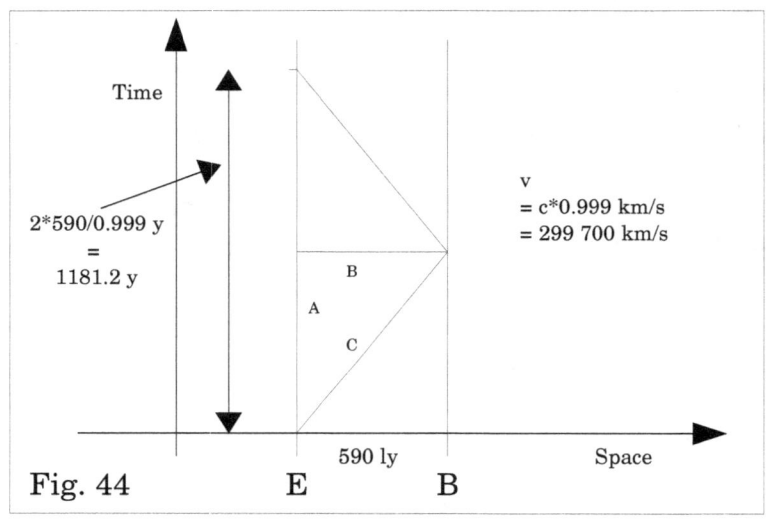

Fig. 44

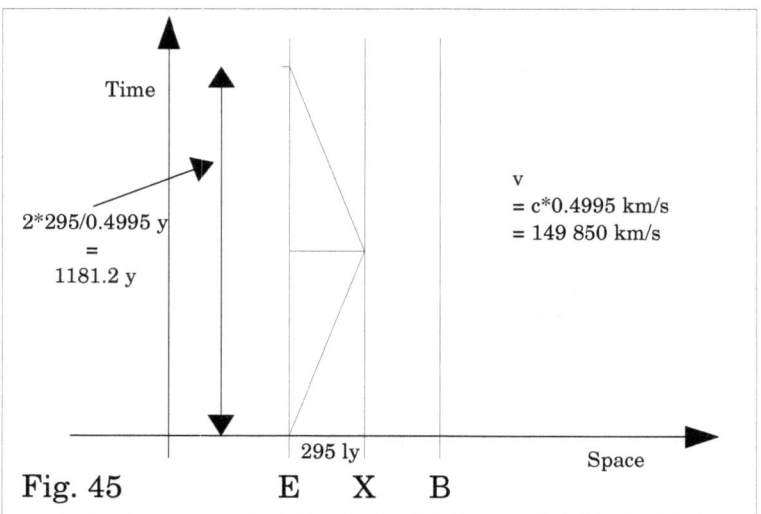

Fig. 45

So in Fig. 44 we show the same thought experiment that was done in the book [14].

The distance between our planet Earth and the star Betelgeuse is 590 lightyears. This is the distance that a light signal will pass for a period of 590 years! Brother $B_2$ moves at a speed of 299 700 km / s.

In Fig. 45 we show thought experiments with brother $B_1$. $B_1$ moves at a speed of 149 850 km / s. That's half the brother's $B_2$ speed. But he'll only fly half the brother's distance.

The two brothers, $B_1$ and $B_2$, return to Earth at the same time. Then it has passed 1181.2 years since they left Earth.

This is logical and corresponds to our way of thinking and calculating with time, speed and distance!

But according to the book [14], the time for the two brothers has passed differently.

When all three brothers are seen again, the following times have passed:
For the brother $B_0$: 1181.2 years
For the brother $B_1$: 1023.3 years
For the brother $B_2$:    52.8 years

All this according to the book [14].

Does anyone believe this?
I'm doubtful!

The author of the book [14], on the other hand, defends these conclusions.
A quote from the book:
"The answer is logical conclusion."

Then we will try to verify this "logical conclusion".
How to do it?

Let's apply the formula for time dilation:
See information from Wikipedia: (swedish)
https://sv.wikipedia.org/wiki/Tidsdilatation
or english
https://en.wikipedia.org/wiki/Time_dilation

$$t' = t\gamma$$

where $\gamma = 1/(1 - v^2/c^2)^{1/2}$ is called the Lorentz factor.
In this formula:
t' = mobile observer's time
t = stationary observer time
c = the speed of light
v = relative speed

We discuss the table below:

| 1 | 2 v (km/s) | 3 c (km/s) | 4 γ | 5 t | 6 t' = tγ | 7 t' = t/γ |
|---|---|---|---|---|---|---|
| Brother $B_0$ | 0 | 300 000 | 1 | 1181,2 | 1181,2 | 1181,2 |
| Brother $B_1$ | 149 850 | 300 000 | 1,15 | 1181,2 | 1363,5 | 1023,3 |
| Brother $B_2$ | 299 700 | 300 000 | 22,37 | 1181,2 | 26419,0 | 52,8 |

In column 4 we have calculated
Lorentz factor $\gamma = 1/(1 - v^2/c^2)^{1/2}$.

In column 5 we indicate the time on Earth, the time during which Brother $B_0$ has been waiting for his two brothers $B_1$ and $B_2$.

In column 6 we have calculated the time according to the formula for time dilation, t' = tγ. But it does not agree with the times calculated according to the formula from the book [14].

In column 7 we have calculated the time according to the formula

t' = t/γ.

These times correspond to the times calculated in the book [14].

Here you can ask the question:
What does time dilatation mean? In which reference system does time dilate?

It is always said that $v$ is the relative velocity between the two reference systems.

This means:
The reference system S moves relatively S' at speed $v$.
The reference system S' moves relative S at speed $v$.
Here we have a symmetrical concept!

How is it possible then that the times in the two reference systems are different?

What about Lorentz transformations?
We should also be able to apply these to calculate the times for the three brothers.

Lorentz transformations:

$$x' = (x - vt)\gamma \qquad \text{(LT}_1\text{)}$$
$$t' = (t - vx/c^2)\gamma \qquad \text{(LT}_2\text{)}$$

where $\gamma = 1/(1 - v^2/c^2)^{1/2}$ is called the Lorentz factor.

We check what values we get for $t'$ according to LT$_2$.

| | 1 | 2 | 3 | 4 | 5 | 6 | 7 |
|---|---|---|---|---|---|---|---|
| | | v (km/s) | c (km/s) | γ | t | x (lightyear) | t' = (t - vx/c$^2$)γ |
| Brother B$_0$ | | 0 | 300 000 | 1 | 1181,2 | 0 | 1181,2 |
| Brother B$_1$ | | 149 850 | 300 000 | 1,15 | 1181,2 | 590 | 1363,5 |
| Brother B$_2$ | | 299 700 | 300 000 | 22,37 | 1181,2 | 1180 | 26419,0 |

We see that the values of $t'$ calculated according to the formula of LT correspond to those calculated by the formula $t' = t\gamma$ (see previous table) but then they do not match those calculated according to the formula $C^2 = A^2 - B^2$ from the book [14].

We conclude this analysis of the twin paradox by showing how to represent grafic quantities time, distance, and velocity in classical physics.

In the classical physics vi have the following formulas:

> length = speed * time → $l = vt$
> time = length / speed → $t = l/v$
> speed = length / time → $v = l/t$

**In classical physics, the diagonal in the right-**

## angled triangle has no meaning.

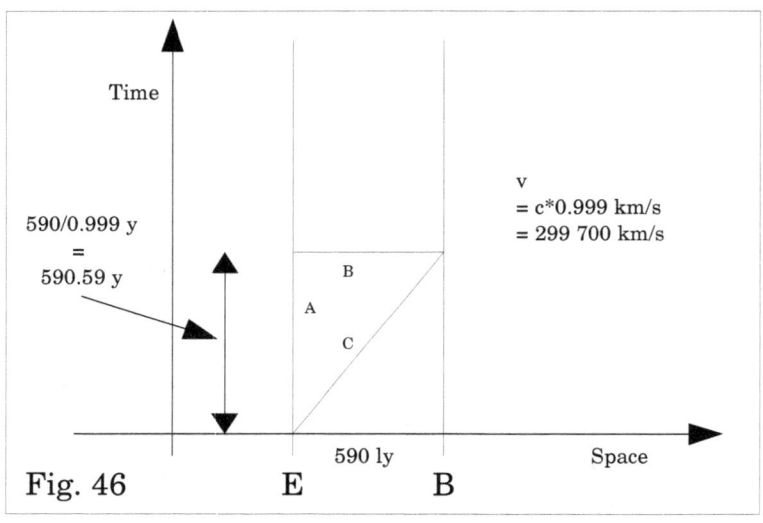

Fig. 46

Here we use the following physical units:
for distance: light year (ly)
for time: year (y)
for speed: light year / year (ly/y)
→
*l = 590 years \*c km / s*
*v = (c km / s)\*0.999*

*t = l / v = (590 years \*c km / s) / ((c km / s)\*0.999 ) =*
*= 590 years / 0.999 = 590.59 years*

It seems that everything is right! Or?

We take another look at the times of the three brothers, times that their clocks show when they are together again on planets Earth.

For the brother $B_0$: 1181.2 years
For the brother $B_1$: 1023.3 years
For the brother $B_2$:    52.8 years

The distance between Earth and planet X is 295 light years.
The distance between planet X and star Betelgeuse is also 295 light years.
Brothers $B_1$ and $B_2$ start simultaneously from Earth and come back at the same time!

During the journey, the distance between $B_0$ and $B_1$ is the same as the distance between $B_1$ and $B_2$.

So why is the time difference between
$B_0$ and $B_1$ 1181.2 years - 1023.3 years = 157.9 years
while between
$B_1$ and $B_2$ it is 1023.3 years -52.8 years = 970.5 years?

This seems to be complete nonsense! See Fig. 47.

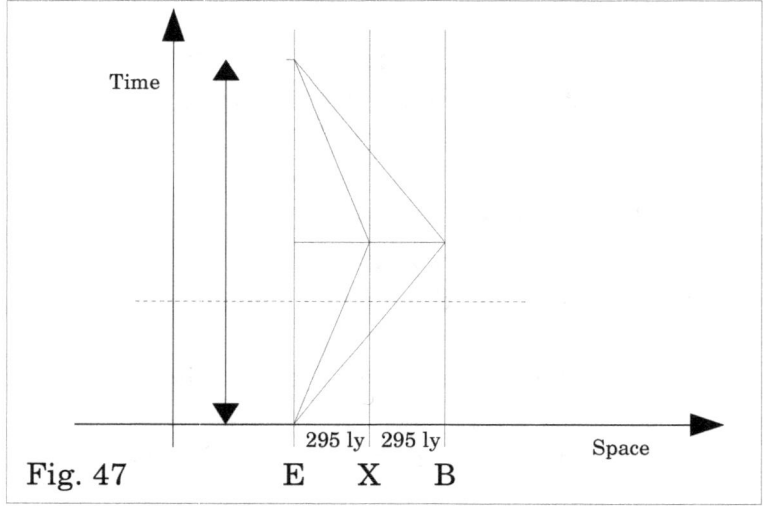

Fig. 47

I let the readers think about these pictures and my calculations and draw their own conclusion.

# My patent application: 1ANM

## 1) 1ANM, Description
## Apparatus for measuring of the absolute velocity in space – 1ANM (One Arm, No Mirrors)

### Background

[0001]
Historical Facts:
- 1864: James Clark Maxwell publishes *Dynamical Theory of the Electromagnetic Field.* Derive that light is an electromagnetic wave. The light should propagate in a medium in the same way that sound waves need air, water waves need water.
This medium was called ether, light-bearing ether.
- 1887: The Michelson-Morley experiment. The purpose of the experiment was to confirm the existence of the ether, measuring the Earth's velocity in space.
- 1904: Hendrik Lorentz formulates Lorentz transformations. This, as a result of that the Michelson-Morley experiment could not demonstrate the existence of the ether.
- 1905: Albert Einstein published the special theory of relativity.

[0002]
The inventor assumes that the Michelson-Morley experiment and the apparatus used, Michelson interferometer, were not suitable for this. The inventor has made his own calculations that show that regardless of the Earth's velocity in space, regardless of the orientation of the apparatus, the difference between these part experiments was in the order of 30 nanometers! This was within the margin of error and therefore the Michelson-Morley experiment got so-called "zero" result, or "negative" result.

[0003]
The inventor believes that the negative result has been caused by the apparatus design: two arms of about 10 meters each, which were mounted at right angles to each other; three mirrors, one of the semi-transparent. The light beam was reflected in these mirrors or passed these mirrors.

[0004]
The inventor believes that the mathematical and physical analysis of the path of light rays was insufficient and in fact incorrect!

[0005]
Therefore, the device (1ANM) is simple and this means that there are no doubts as to how the light moves inside the device (1ANM). Even the mathematical and physical analysis of the path of the light beam becomes simple, as is the calculation and measurement of the

absolute velocity in space becomes easier and clearer!

## Description

[0006]
*The device (1ANM) is based on the principle that the speed and direction of light are independent of the movements of the source and the observer.*

[0007]
List of Figures
Fig. 1 - the device (1ANM), consisting of an arm (A), a laser (L), a screen (S)
Fig. 2 - a section through the apparatus (1ANM); the screen (S) is marked with a coordinate system (x-axis, y-axis); the line connecting the laser (L) to the center of the screen (S) represents the third axis (z-axis) of a three-dimensional coordinate system.
Fig. 3 - two simpler representations of the device (1ANM) that will be used in the various measurements / calculations: a dotted line image and a full line image. On the right side of the image, there is depicted vector *v* for the absolute speed.
Fig. 4 - shows two possible directions for the absolute speed *v*
Fig. 5 - compilation of the results of eight different measurements / calculations

Fig. 6 - shows how to determine the direction of the absolute speed vector

[0008]
Fig. M1 – measurement / calculation 1
Fig. M2 – measurement / calculation 2
Fig. M3 – measurement / calculation 3
Fig. M4 – measurement / calculation 4
Fig. M5 – measurement / calculation 5
Fig. M6 – measurement / calculation 6
Fig. M7 – measurement / calculation 7
Fig. M8 – measurement / calculation 8

[0009]
The device (1ANM) consists of the following parts, see Figs. 1 and 2:
- an arm (A) on which the other parts are mounted
- a laser (L) that creates a light dot (P) (with about 1 millimeter in diameter)
- a screen (S) on which the dot (P) ends and where the position of the dot (P) can be read in relation to the center of the screen (S)

[0010]
The arm (A) must be attached to a device (similar to a gyroscope or a robot arm) that allows the arm (A) of the device (1ANM) to be directed, rotated in all possible directions in space.

This part is not part of our invention. There are various technical possibilities to solve this.
[0011]
The distance between the laser heads (L) and the screen (S) must be at least 10 meters.

[0012]
Detailed description of the device (1ANM) components.

[0013]
Arm (A):
At one end of the arm (A) a laser (L) is mounted. At the other end of the arm (A) a screen (S) is mounted. The length of the arm (A) should be such that after mounting the laser (L) and the screen (S), the distance between the laser (L) head and the screen (S) should be at least 10 meters. The arm (A) should be of square cut, and should be designed so firm and stable that no bending of the arm occurs. It is not essential in the context of which material the arm (A) is made.

[0014]
Laser (L):
Laser (L) should be able to transmit a light beam, which is imaged on the screen (S) as a dot (P) of about 1 millimeter in diameter. It is not essential in the context which type of laser is used.

[0015]
Screen (S):
The screen (S) should allow the dot (P) position to be read. The screen (S) should be of square shape, preferably giving the possibility to digitally read the dot (P) position in the coordinate system (x, y) and send this position to a computer for further processing. It is not essential how the screen (S) is manufactured. The size of the screen (S) (readable part) should be at least 6x6 centimeters. This ensures that the light dot (P) movement on the screen (S) does not fall outside. The maximum calculated speed is approx. 850 km / s. This corresponds to the maximum distance of the dot (P) to the center of the screen (S) to about 3 centimeters.

[0016]
Measurements / Calculations:

[0017]
We will use the following terms:
- $D$ = the length between the laser (L) and the screen (S) (10 meters)
- $c$ = light speed (300,000 km/s = 300,000,000 meters / second)
- $v$ = the speed at which the Earth moves in space and with that also the device (1ANM)
(eg 30 km/s = 30,000 meters/second); it is the speed at

which the point on the Earth's surface moves, the point at which the apparatus (1ANM) is located; ***it is this speed that will be read / calculated using the device (1ANM)***
- $t$ = the time the laser beam needs to reach the screen (S)
- $d$ = distance between the dot (P) and the center of the screen (S)
- $x$ = in Fig. M1-M8, $x$ is a variable used in the calculation
(should not be confused with x-coordinate)

[0018]
We present eight part experiments/measurements / calculations to show how the device (1ANM) moves in space, how the laser beam moves. We make calculations to show where somewhere on the screen (S) the light dot (P) ends up depending on the device's (1ANM) orientation in space and its speed $v$.


[0019]
These 8 part experiments are shown in Fig. M1-M8. In the upper part we show the original position, depicted with dotted line. Below, there are two overlapping (partially overlapping) images on the device (1ANM), one with dotted lines for the initial

position and one with full line for end position, then when the laser beam reaches the screen (S).

[0020]
To the right of the device (1ANM) is shown the vector for the speed $v$ (which direction $v$ has and which orientation the apparatus (1ANM) has against this vector).

[0021]
A point on the Earth's surface has the following movements:
1) Movement around the Earth's axis; at equator, the speed is approx. 0.5 km / s
2) Earth's movement around the Sun; the speed is approx. 29.8 km / s
3) The movement of the solar system around the center of the galaxy; the speed is approx. 220 km / s
4) The galaxy moves toward the Great Attractor; the speed is approx. 600 km / s
5) The big attractor, in turn, moves toward the Shapleys superhope, which is a collection of over 8,000 galaxies (not included in my calculations).

[0022]
This means that the absolute speed can be at least approx. 350 km / s
(600-220-30).

With this speed will be ***max(d) = 11 millimeter***.

[0023]
An example of measurement / calculation:
We measure the maximum distance between position of the light dot (P) and the center of the screen (S) ***d***
d = 7 millimeter = 0,007 meter
v = cd/L = 300 000 000 m/s * 0,007 m / 10 m
v = 210 000 m/s = 210 km/s

[0024]
We now present eight measurements / calculations corresponding to eight different relative positions of the apparatus (1ANM) arm (A) with respect to the vector of speed ***v***.
These part experiments / measurements / calculations predict that the apparatus (1ANM) is positioned such that y-axis, z-axis and vector $v$ are in the same plane and that when the apparatus (1ANM) rotates 360° in this plane, around the x-axis .

[0025]
*Measurement / calculation 1, Fig. M1:*
*The device (1ANM) moves parallel to the vector **v**, to the right.*
*While the laser beam reaches the screen (S), the entire apparatus (1ANM) moves with distance **x** to the right.*

*In this case, the laser beam needs to pass distance D-x.*

*The light dot (P) ends up exactly in the center of the screen (S).*

$t = x/v = (D-x)/c$
$x = Dv/(c+v)$
$d = 0$

[0026]
*Measurement / calculation 2, Fig. M2:*

*The device (1ANM) moves to the right and upwards and forms an angle of 45° with the vector **v**.*
*The speed is $v_1 = v/2^{1/2}$ (root of 2).*

*While the laser beam reaches the screen (S), the entire apparatus (1ANM) moves by distance **x** to the right and by the same distance **x** upwards.*
*In this case, the laser beam needs to pass the distance **D-x**.*

*The dot (P) falls down on the center of the screen (S).*

$t = x/v_1 = (D-x)/c$
$x = Dv_1/(c+v_1)$
$d = x$

[0027]
*Measurement / calculation 3, Fig. M3:*

*The device (1ANM) moves upwards and forms an angle of 90 ° with the vector **v**.*
*While the laser beam reaches the screen (S), the entire apparatus (1ANM) moves with distance **x** upwards.*
*In this case, the laser beam needs to pass the distance D.*

The dot (P) falls down on the center of the screen (S).

$t = x/v = D/c$
$x = Dv/c$
$d = x$

[0028]
*Measurement / calculation 4, Fig. M4:*

*The device (1ANM) moves up and to the left and forms an angle of 45° with vector **v**.*
*The speed is $\boldsymbol{v}_1 = v/2^{1/2}$ (root of 2).*
*While the laser beam reaches the screen (S), the entire apparatus (1ANM) moves by distance **x** upwards and by the same distance **x** to the left.*
*In this case, the laser beam needs to pass the distance **D+x**.*

The dot (P) falls down on the center of the screen (S).

$t = x/v_1 = (D+x)/c$
$x = Dv_1(c-v_1)$
$d = x$

[0029]
*Measurement/calculation 5, Fig. M5:*

*The device (1ANM) moves left and parallel to vector **v**. While the laser beam reaches the screen (S), the entire device (1ANM) moves distance **x** to the left.*
*In this case, the laser beam needs to pass the distance **D+x**.*

*The light dot (P) ends up exactly in the center of the screen (S).*
$t = x/v = (D+x)/c$
$x = Dv(c-v)$
$d = 0$

[0030]
*Measurement/calculation 6, Fig. M6:*

*The device (1ANM) moves left and down and forms an angle of 45° with vector **v**.*

*The speed is $v_1 = v/2^{1/2}$ ( root of 2).*

*While the laser beam reaches the screen (S), the entire apparatus (1ANM) moves with distance $x$ to the left and with the same distance $x$ down.*
*In this case, the laser beam needs to cut off the distance $D+x$.*

*The light dot (P) ends up on the center of the screen (S).*

$t = x/v_1 = (D+x)/c$
$x = Dv_1(c-v_1)$
$d = x$

[0031]
*Measurement/calculation 7, Fig. M7:*

*The device (1ANM) moves downwards and forms an angle of 90° with the vector $v$.*
*While the laser beam reaches the screen (S), the entire apparatus (1ANM) moves with distance $x$ downwards.*
*In this case, the laser beam needs to pass the distance $D$.*

*The light dot (P) ends up on the center of the screen (S).*

$t = x/v = D/c$
$x = Dv/c$
$d = x$

[0032]
*Measurement / calculation 8, Fig. M8:*

*The device (1ANM) moves to the right and down and forms an angle of 45° with the vector **v**.*
*The speed is $v_1 = v/2^{1/2}$ (root of 2).*
*While the laser beam reaches the screen (S), the entire apparatus (1ANM) moves with distance **x** to the right and with the same distance **x** down.*
*In this case, the laser beam needs to pass the distance **D-x**.*

*The light dot (P) ends up on the center of the screen (S).*

$t = x/v_1 = (D-x)/c$
$x = Dv_1(c+v_1)$
$d = x$

[0033]
In Fig. 5 we show a summary of the above eight experiments/measurements/calculations and over the position of the light dot (P) on the screen (S) during a

complete rotation of the apparatus (1ANM) by **360°**.
The calculations show that the distance of the light dot (P) to the center of the screen (S), **$d$**,
would vary between 0 millimeters and 1 millimeter (if **$v = 30\ km\ /\ s$**).

[0034]
Fig. 6 shows how the direction of **the absolute speed** is determined.
After finding the maximum distance **$d$** that the dot (P) has towards the center of the screen (S), the vector between position of the dot (P) and the center of the screen (S) determines the direction of **the absolute velocity** in space.

[0035]
In this way, one can determine both the **scalar value** **$v$** for the absolute velocity in space and its **direction**.

## 2) 1ANM, Claim

# CLAIM

**Apparatus for measuring the absolute velocity in space -
1ANM (One Arm, No Mirrors)**

**What I claim as my invention is:**

**Claim 1.
An apparatus for measuring the absolute velocity in space. The apparatus consists of an arm (A), a laser (L) and a screen (S) arranged so that the laser (L) is mounted at one end of the arm (A) and the screen (S) is mounted at the other end of the arm (A ). Laser (L) transmits a laser beam to the screen (S) where it is projected as a dot (P). The position of the dot (P) is read against the center of the screen (S).**

*The device (1ANM) is mounted on a device that allows the arm (A) to be rotated in all possible directions in space. This device is not included in the patent claim because there may be several technical solutions to achieve this (robot arm; gyroscope).*

The device (1ANM) is rotated until the y-axis and z-axis land in the same plane as the vector for speed $v$ (Earth's velocity in space). The method by which this is achieved is not included in the claim because it is dependent on the device on which the arm (A) is mounted.

**Claim 2.**
**This claim consists of the description of the exact path of light through the apparatus, the calculations of the scalar absolute speed, the method of determining the direction of absolute velocity in space. The apparatus (1ANM) together with the description, calculations and measurement method represents the simplest apparatus for measuring of the absolute velocity in space and its direction.**

When the apparatus's y-axis, z-axis and the velocity vector $v$ are in the same plane, the measuring method is applied, which is **characterized** in that the arm (A) is rotated 360° in the same plane and that the maximum distance $d$ between the laser beam dot (P) and the center of the screen (S). The greatest distance is obtained when the arm (A), z-axis, constitutes an angle of 90° to the vector $v$ of the velocity. Then the calculation is **characterized** by the following formula

$$v = dc/D$$

where
- $v$ is the absolute speed at which the Earth moves in space (the point on Earth where the apparatus is)
- $d$ is the largest distance between the position of the dot (P) and the center of the screen (S)
- $c$ is the speed of light (in this case in air)
- $D$ is the distance between the laser head (L) and the screen (S)

## 3) 1ANM, Summary

## SUMMARY

**Apparatus for measuring the absolute velocity in space -
1ANM (One Arm, No Mirrors)**

Scope: Physics.

*The device (1ANM) is based on the principle that the speed and direction of light are independent of the movements of the source and the observer.*

The device (1ANM) consists of an arm (A), a laser (L) and a screen (S). The device (1ANM) is mounted on a device that allows the arm (A) to be rotated in all possible directions in space. Laser (L) transmits a laser beam to the screen (S) where it is projected as a dot.

The **value of the absolute speed $v$** is calculated using the maximum distance $d$ between the position of the dot (P) and the center of the screen (S), which is read at a full rotation of the apparatus (1ANM) by 360°.

This value is calculated by the formula $v = dc/D$.

The **direction of the absolute velocity** $v$ in space is determined by the vector created between the light dot (P) and the center of the screen (S) and read on the y-axis at which the reading of the maximum distance of the dot (P) to the center of the screen (S) was made.

## 4) 1ANM, Drawings

Presented on the following 6 pages.

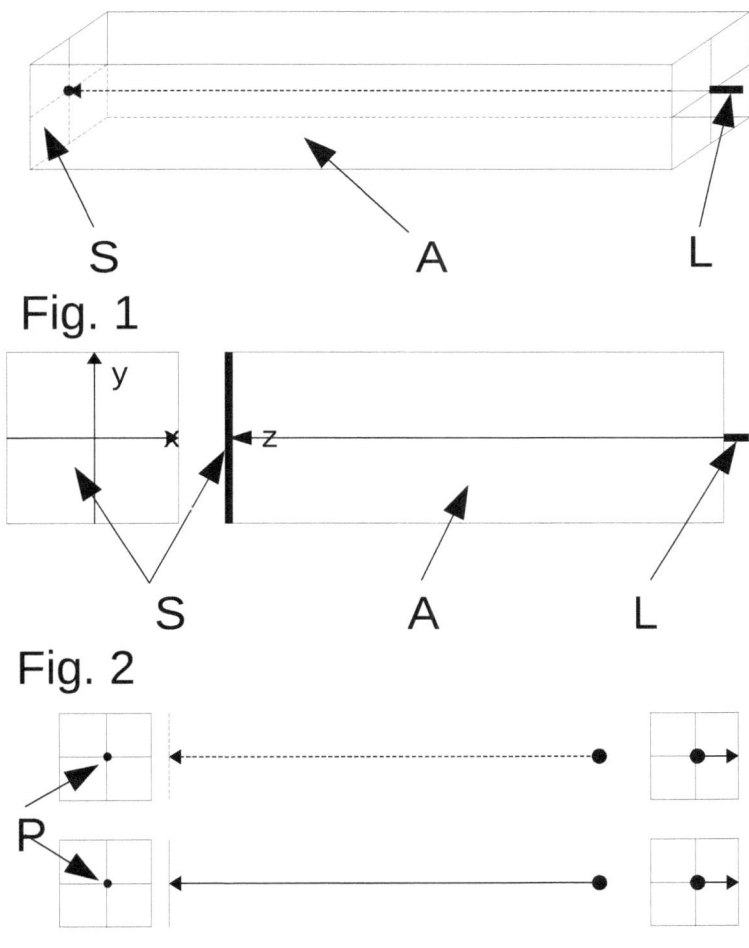

Fig. 1

Fig. 2

Fig. 3

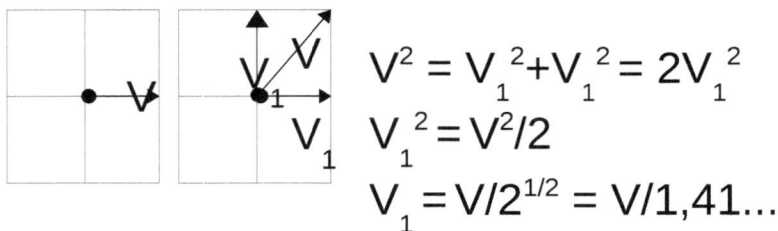

Fig. 4

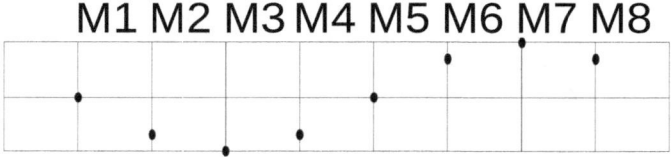

Fig. 5

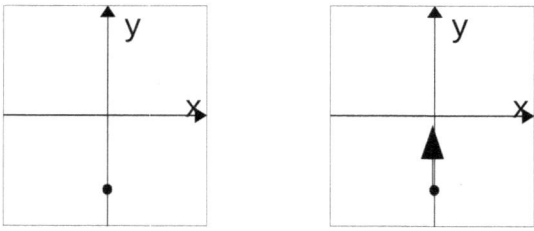

Fig. 6

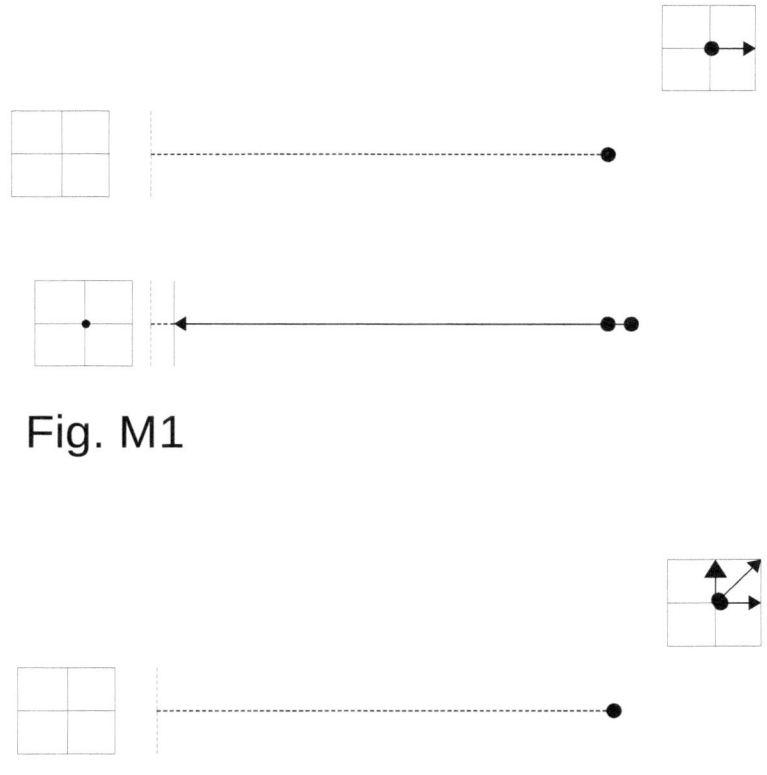

**Fig. M1**

**Fig. M2**

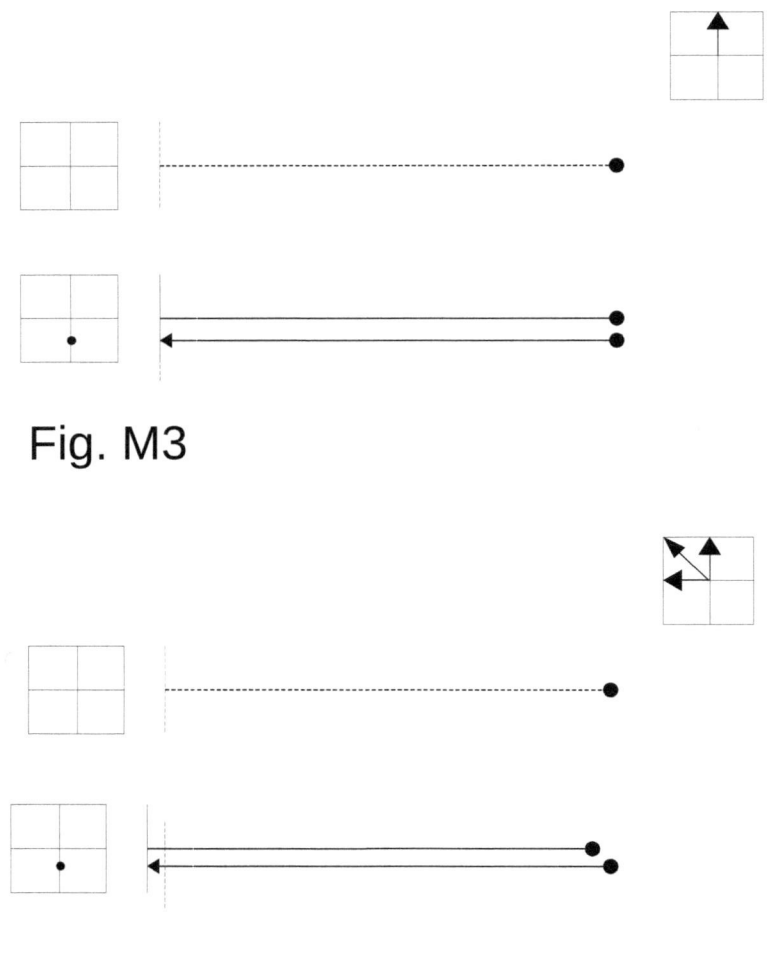

Fig. M3

Fig. M4

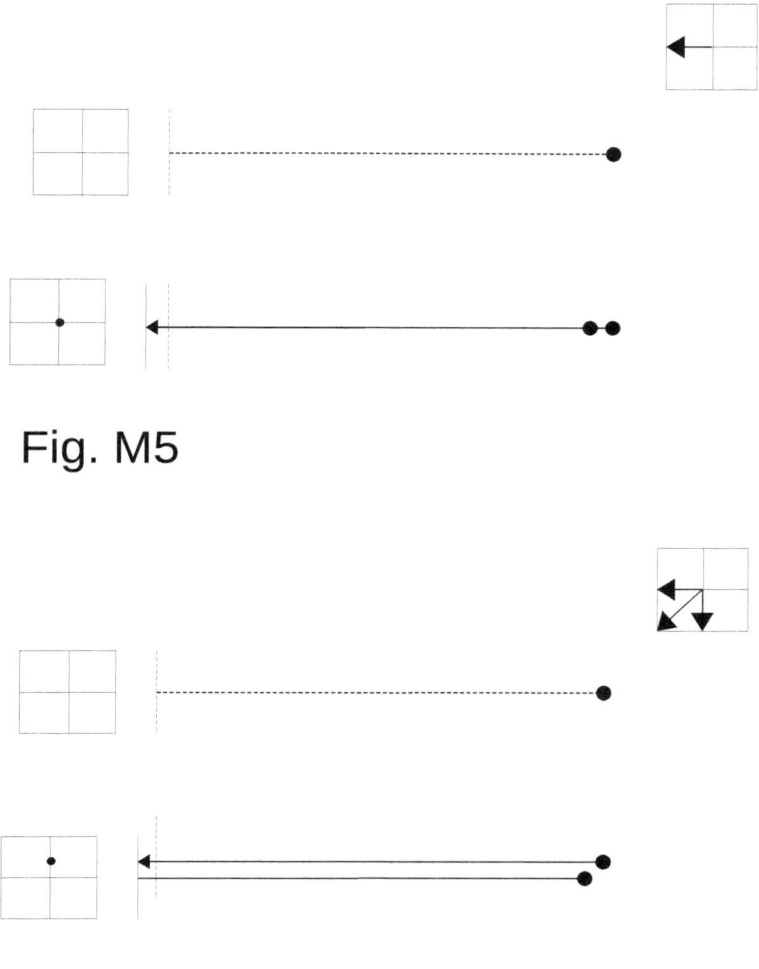

Fig. M5

Fig. M6

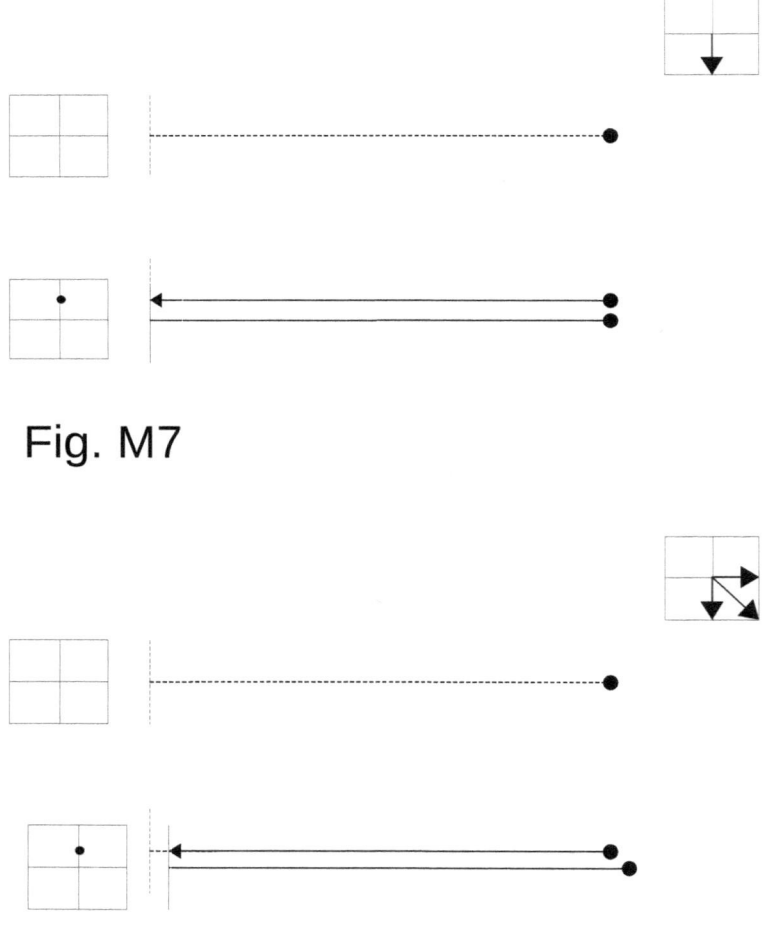

**Fig. M7**

**Fig. M8**

# Ending

In this work, we analyzed the following:

A1) Experiments with two reference systems, S and S', stationary against each other, and an object in S in which events occur.

A2) Experiments with two reference systems, S and S', moving at constant speed $v > 0$ against each other, and an object in S' in which events occur.

**We show that the same coordinate transformation exists either the two reference systems, S and S', are stationary towards each other or if they move at a constant speed $v > 0$ with respect to each other.**

A3) Time dilation in *[1], [6] and [8]*.

**We show how completely wrong one reasoned about the propagation of light. We show that 'light clock' works in the same way in two reference systems, S and S', which move at constant speed $v > 0$ with respect to each other. Thus, we show that no time dilation occurs in any of the two reference systems!**

A4) Derivation of Lorentz transformations in *[7]*.

**We show here that the calculations are incomplete and that one comes into contradiction with the conditions of origin that make this derivation incorrect!**

A5) Derivation of Lorentz transformations in *[3]*.

**We show how this derivation is based on incorrect mathematical assumptions and that one comes into contradiction with the conditions of origin that mean that even this derivation is incorrect!**

A6) Michelson-Morley experiment, 1887

**The theoretical preparation of this experiment was insufficient! One has not made a complete analysis of the path of the light beams through the interferometer and therefore erroneously the magnitude of the changes in the interference pattern.**

**We show by means of a theoretical analysis and calculations that the changes in the interference**

**pattern that occurred by turning the interferometer by 90° were extremely small, which was probably impossible to measure at that time.**

A6) Other chapters ...

**These show how I perceive the special theory of relativity. Here I come with some new arguments which everyone leads to contradictions, nonsense.**

**My analysis of the special theory of relativity shows too many errors, too many misinterpretations!**

**From this, one should conclude that the special theory of relativity is incorrect in its entirety!**

> The special theory of relativity
> is a great nonsense!

# Quotes from books I read
and my comments (sometimes) with capital letters

## 1)
### *Einstein, chaos and black holes; K. E. Cole;*

*p37:*
"The simpler the models, the longer they are from reality. Still, the simplest models are often the most useful.
This is one of the reasons why mathematics is such a useful tool in physics. It is the ultimate abstraction ... "
"The imagery lets us go faster, but the truth lies in mathematics ..."
"Today, mathematics has to a large extent become the language of science."

*p38:*
"And in physics one has made as many discoveries by looking at equations as by looking in the microscope and telescope"
Feynman: "It is almost something mysterious that mathematical thinking seems to get things together"

*p48:*
"In the long run, it will almost certainly prove that Einstein was wrong. At least in the same sense that he

himself showed that Newton was wrong. "

*p52:*
James Jeans: "In real science, a hypothesis can never be proven. If it is refuted by future observations, we know that it is wrong, but if future observations confirm it, we can never say that it is right, because it is always extradited on mercy and evil for new observations "

*p155:*
"The general theory of relativity arose when Einstein expanded the thoughts of the special theory of relativity to apply to all kinds of movement - especially the changing or accelerating movement of objects falling under the influence of gravity."

## 2)
## *Black holes; Bengt Gustafsson;*

*p81:*
"Einstein was not the first to work with the theory, and he learned a lot by studying and partly copying the efforts of the others"

*p101:*
"One of the fundamental problems he had not solved was how Lorentz transformations, which have been so

successful in the special theory, could be applied in the field of gravity, if it was possible at all"

## 3)
## *ABC of Relativity; B. Russell; 1925;*

*p321:*
Novikov; 1980; Self Consistency principle;
"According to this, time travel that gives rise to violation against causality is impossible ... a rule for all theory building - theories in physics simply must not allow causality violation"

## 4)
## *At the crossroads; Ulf Sinnerstad;*

*p58:*
"At the same time, it is these worlds that we have the most difficult to understand, worlds where obvious models pass us and we find paradoxes. But there is only one world and it has no paradoxes. Only our models have paradoxes "

*p196:*
"If we measure the time-space coordinates - x, y, z - in length measure, the time coordinate t must also be given in length measure

$$s^2 = c^2t^2 - (x^2 + y^2 + z^2)\text{ ''}$$

## 5)
### *Cosmos - a short story; Stephen Hawking;*

*p15:*
"A physical theory is always provisional in the sense that it is merely a hypothesis: one can never prove it. However many times the experimental results are consistent with a particular theory, one can never be sure that the results next will contradict the theory. On the other hand, one can disprove a theory by finding only one that does not conform to the predictions of theory "

*p17:*
"Newton's theory also has the great advantage of being much easier to work with than Einstein's"

*p18:*
"Today, the researchers describe the universe with the help of two basic sub-theories: the general theory of relativity and quantum mechanics.
Unfortunately, however, we know that these two theories are incompatible - they cannot both be true "

*p35:*
"It is impossible to imagine a four-dimensional space"

## 6)
## *An exquisite universe; Brian Green;*

*p50:*
"Imagine you are on a train and pull down the blinds so that the windows are completely covered. If you have no opportunity to see anything outside your own compartment and provided that the train moves at a completely constant speed, there will be no way for you to determine your movement. The cabin around you will look exactly the same regardless of whether the train is standing still on the tracks or moving at high speed. Einstein formalized this idea, an idea that actually goes back to Galilei's insights, claiming that it is impossible for you or any of your fellow passengers to carry out any experiment in the closed compartment that can determine whether the train is moving or not ...
There is no way for you to decide anything about your state of motion without making direct or indirect comparisons with 'outer' objects. There is simply no such thing as 'absolute' movement at constant speed; only comparisons have physical meaning "

WRONG! YOU CAN DECIDE IF YOU ARE MOVING OR NOT!

*p54:*
"Regardless of the relative motion of the photon source and the observer, the light speed is always the same"

"But this triumph over the conflict was no small victory. Einstein realized that the constancy of lightning meant the fall of Newtonian physics"


*p57: train*
"Since the speed of the light emitted to the left or to the right is the same, they consider - and actually observed - that the light apparently reached both presidents at the same time. Who's right?
Those on the train or those outside?
Each group's observers and explanations in support of them are immeasurable.
The answer is that both are right!"

WRONG! ONLY ONE CAN BE RIGHT!

*p60:*
"Our goal is to understand how movement affects the passage of time, and since we defined the time operationally by means of clocks, we can translate our question into how movement affects the clocks' ticking"

*p61:*
"The reason we use a light clock in our discussion is that its simplicity in mechanical respect scales off all superfluous details and therefore gives us the clearest insight into how movement affects the passage of time"
...
"The question we ask is whether the light clock in motion ticks at the same rate as the light clock at rest?"
...
"The photon starts at the bottom of the sliding (moving) clock ... and travels first to the upper mirror. As the watch moves from our perspective, the photon must move at an angle "

WRONG! THAT IS THE MOST DUMB I HAVE READ IN SCIENCE !!!
THE UNIVERSE DO NOT WORK AFTER SR!
...
"If the photon did not move along this path, it would miss the upper mirror and go into space"
SO WHAT?!

*p67:*
"Like all apparent paradoxes arising from the SR, these logical dilemmas are dissolved by closer scrutiny and reveal new insights into how the universe works"

# 7)
## *The dust of which the cosmos is woven; Brian Green;*

*p18:*
"For Gottfried Wilhelm von Leibniz, 'space' and 'time' were simply words for connection between where objects were and when events occurred ..."

*p50:*
Mach: "In an otherwise empty universe, it is not possible to distinguish between standing completely motionless and rotating uniformly"
"... if your body rotates ... every part of your body rotates at the same rate"

WRONG! YOUR AXLES HAVE HIGHER ROTATION SPEED THAN YOUR NECK!

*p63:*
"But the light speed is constant; space and time behave in this way. Space and time adjust in a way so that they exactly compensate for each other, so that observations of the light travel give the same result regardless of the observer's speed "

*p72:*

"Einstein's unexpected answer is that both are right. Although the two judges' conclusions differ, each of the judges' observations and reasoning is inviolable "

WRONG! ONLY ONE HAS RIGHT!

*p76:*
"So, although Newton was definitely wrong, the special theory of relativity did not completely crush his intuition that there is something absolute, something everyone would agree on.
Absolute room does not exist.
Absolute time does not exist.
But according to the special theory of relativity, absolute spacetime exists.
...
In order for a subject's path through spacetime to be a right line, the object must not only move in a right line through the room, but its movement must also be uniform throughout time, that is, its speed and its direction must be unchanging and thus it must moves at a constant speed."

*p95:*
"Everyday experience therefore fails to reveal how the universe actually works, and that is why even one hundred years after Einstein almost no one, not even a physicist, has the theory of relativity in the spinal

cord"

p96:
"The special and general theory of relativity pointed out important spikes in the clockwork image: there is no single universal clock that has precedence"

**YES, IT DOES IT! Light clock!**

p187:
"The centuries of scientific research have shown that mathematics is a powerful and fierce language with which we can analyze the universe. ...
Physicists have come to realize that mathematics when used with sufficient caution is a proven path to truth "

p255:
"No one has so far found the definitive, basic definition of time, but undoubtedly part of the role of time in the cosmic construction is that it accounts for changes. We notice that the time has passed by observing that things are different now to what they were before. "

**8)**
***A brief history of almost everything; Bill Bryson;***

*p115: about ether*
"Decartes had presented the idea, Newton had supported it and then almost everyone had praised it. The ether was absolutely central in the 19th century physics as a way of explaining how light could travel through the emptiness of space. It was especially necessary in the 19th century because light and electromagnetism were then considered as waves, that is, a kind of vibration. Vibrations must take place in something; therefore, they needed and held fast to the ether. As recently as 1909, the great British physicist J. J. Thompson maintained: "The Ether is not a fantastic creation of a spectacular physicist; it is as important to us as the air we breathe"

*p116:*
"In fact, it was understood, of course, that the world was entering a century of science when many people would not understand anything and no one would understand everything"


**9)**
***In the head of God; Paul Davies;***

*p25:*
"The claim that the world is rational is linked to the fact that it is orderly. Events generally do not happen

anyhow: they are related in some way. ... It is this relationship of events that gives us our perception of cause and effect. (Causality)
...
The concept of determinism is closely related to causality. In its modern form, this is the assumption that events are entirely determined by other, past events. "

*p66:*
"As long as the universe had a beginning, we could assume it had a creator ..."

*p93:*
"There is no subject that better illustrates the difference between... humanities and natural science than mathematics. For the outsider, mathematics is a strange, abstract world of horrible technicalities, full of bizarre symbols and complicated procedures, an impenetrable language and a black magic. For the natural scientist, mathematics is the guarantee of accuracy and objectivity. It is also surprisingly enough the nature's own language. Nobody excluded from mathematics can ever grasp the whole meaning of the order of nature so deeply embedded in the fabric of physical reality.
...
But the notion that mathematics is a key that gives

the inserted opportunity to unlock cosmic secrets is as old as the subject itself "

## 10)
### *Stars and apples falling; Ulf Danielsson;*

*p134:*
"Thinking freely is great, but thinking right is greater. Thomas Thorild, Uppsala University.
...
It is better to think wrong than to not think at all. Hypatia, Alexandria "

*p135:*
"And who knows, maybe the story has more dark periods in readiness in the future? There are no guarantees that free thinking and philosophy will persist "

## 11)
### *Big bang or let there be light?; Maria Gunther Axelsson;*

*p30-31:*
"Therefore, the research is not about defending established theories (which the creationists often claim), but rather testing, questioning and trying to

find gaps and inaccuracies in the theories"
*p32:*
"A scientific theory can only be rejected if it makes false predictions"

**I do this in my book, in my analysis of SR!**

*p69:*
"For most of us, it is hardly enough with a basic education in physics to, for example, understand Einstein's theory of relativity (I have studied technical physics at university and then a doctorate in particle physics, but I do not belong to those who seriously understand the theory even if I can use it and count on it). "

*p88:*
"No one today believes in ... ether (the mystical medium that would fill the universe so that light could arrive). The theories have been replaced by better models for how the world works. When the physicists in the end of the 19th century had to realize that the ether does NOT exist (after some famous experiments), they needed a better theory of light (and they soon received it in Einstein's special theory of relativity).

**12)**

## *The big plan; Stephen Hawking;*
*p33:*
"Our book is based on the scientific determinism, which means that the answer to question two becomes: There are no miracles or other exceptions from the natural laws"

*p39: model-dependent realism*
"... the idea that a physical theory or world view is a model (usually mathematically formulated) with a set of rules that bind the model's various elements or details to observations."

*p55:*
"There seems to be no mathematical model or theory that alone can describe the universe from all aspects."

*p81:*
"The world is understandable because it is governed by scientific laws, that is, you can express how it works with the help of mathematical models"

*p87:*
"When Maxwell said he had discovered that the" light speed "looked out of his equations, ... the natural question was what the light speed in his equations was measured relative to .

...

that his equations indicate the light speed in relation to an unknown medium that fulfills the whole space, the light-bearing ether

...

If the ether existed, there would be an absolutely immobile state (that is, the ether) and consequently an absolute definition of motion. The ether would provide a suitable reference system that included the entire universe and against which one could measure the speed of all objects. The ether was suggested from theoretical considerations, and some physicists hastened to try to find ways to investigate it or at least confirm that it existed. "

## 13)
### *The heavy science; Bo Dahlin;*

"Researching is to think"

"Science is a special way of knowing and investigating and the only way of appreciating the process is to do it. Only in this way can people came to recognize a key feature of science: there is only one correct explanation for any set of phenomena. Finding that correct explanation can be difficult, painful, ..., frustrating, fun, and ultimately very reewarding"

## 14)
## *Modern Physics; Kenneth Krane;*

*p20:*
"This theory has a completly undeserved reputation as being so exotic that few people can understand it"
"the special theory of relativity has been carefully and thoroughly tested by experiment and found to be correct in all its predictions"

## 15)
## *Understanding Physics; M. Mansfield, C. Sullivan;*

*p193:*
"the velocity of light is invariant ... this result is in direct conflict with the findings of the previous chapter where the Galilean tarnsformations showed us that the value of the velocity of an object must change when it is measured from a coordinate system which is moving"

*p194:*
"Michelson and Morley found no difference between the value of $c$ measured in referens system moving parallel to and perpendicular to the Earth's motion and hence found no evidence of the existence of an

ether"
**16)**
***University Physics; Young, Freedmann;***

*p1243:*
"Einstein's conceptual leap was to recognize that if Maxwell's equations are valid in all inertial frames, then the speed of light in vacuum should be the same in all frames and in all directions"

An observer at rest against the light source and one in motion must measure the same light speed, $c$.

BUT THEY NEVER SAY HOW TO MEASURE THE LIGHT SPEED!

**17)**
***Beyond Reason; A. K. Dewdney; 2004;***
*p4:*
" perpetual motion machine: it is not suprising that more than a few had faked the results … But all of them failed"

*p5:*
"it was not until the mid nineteenth century that we finally understod that the project was doomed. (the teory of thermodynamics)"

*p6:*
"there are mathematical steps that invoke the grander edifice lurking in the background"

*p7:*
"... and relativity theory are practically all mathematics ... It was Kuhn who argued that scientific revolutions have their roots not so much in data, but in how we interpret them"

*p45:*
"Michelson and Morley set up the interferometer in what they thought might be the direction of the Earth travel through ether, but found that the interference fringe did not shift in their instrument. Well, perhaps, they hadn't guessed right, so they tried another angle. That produce no effect, either. Their tried every angle they could think of, even tillting the interferometer tovard the ceiling. Still no effect. They conclude that there was no ether, at least not one with the properties attributed to it. In short, light was not propagate through any misterious medium. Moreover, the speed seemed to be the same in all directions!"

*p46: Fitzgerald*
"According to the outcome of the Michelson-Morley experiment, light always traveled at the same speed, regardless of the state of motion of it source.

... He did not belive that the velocity of light could be unaffected by the motion of its source. The only escape from this logical cul-de-sac, as far as Fitzgerald was concernde, was to suppose that any object in a state of motion was subject to a contraction in the direction of its motion."

## 18)
## *Nothing; Frank Close;*

*p59: The problem of the ether*
"Drop a stone into water and a wave spreads out. The speed of the wave is about a meter per second. This speed is a property of the water. It does not depend on the velocity of the source. If the stone is dropped in from a stationary boat, the waves spread at 1 meter each second; if dropped in from a speedboat they still spread at 1 meter per second. If you are on a boat that is at rest in the water, you will see the wave pass you at a speed of 1 meter each second. If however you were heading into the waves at 10 meters per second the waves wolud approach you at 11 meters per second, whereas if you were headed the other way at the same speed relativ to the water, you would be overtaking the wave at 9 meters per second. You can determine your absolute speed relativ to the water this way. As it was the boat in the water, so it would be for the Earth in the ether."

## LAST SENTENCE: POSSIBLY WRONG!

*p63:*
" Today we know that velocity dependent transformations are correct, length do contract and masses do grow with increasing speed in proportion to the factor
*$1/(1 - v^2/c^2)^{1/2}$* but not for reasons that Lorentz and Fitzgerald had suggested. Einstein took a new perspective on the problem.
The invariance of the speed of light with respect to the speed of source or observer is a result, in part, of distances contracting as in Lorentz and Fitzgerald's formula but this was not due to any ether acting on the rod. For Einstein the contraction are an intrinsic property of spaces itself."

*p65:*
"The fact that the velocity of light is independent of the speed both of the source and of the receiver was an enigma."

*p68:*
Speed is a measure of the distance travelled in an interval of time (definition)

"Speed is the ratio of distance moved to time elapsed and relative speeds add or subtract depending on whether you are heading towars or are running away from a speading object. However, common sense feils for light beams, since independent of how fast you move or in what direction, your relative speed to a light beam is invariant. Einstein realized that something must be wrong with our concept of space and time.

**19)**
***Introduction to Physics; J. D. Cutnell, K. W. Johnson;***

*p4:*
"Only quantities with the same units can be added or substracted."

*p30:*
"Definition of average velocity:
Average velocity = Displacement/Elapsed time"

*p735: 1865, Maxwell*
"determined theoretically that electromagnetic waves propagate through a vacuum at a speed given by
$c = 1/(\varepsilon_0 \mu_0)^{1/2}$

*p876:*
"Since the laws of physics are the same in all referens frames, there is no experiment that can distinguish between an inertial frame that is at rest and one that is moving at a constant velocity"

YES, THERE IS AN EXPERIMENT !!!

*p877:*
"According to this view, an observer moving relative to the ether would measure a speed of light that was smaller or greather than $c$, depending on whether the observer moved with or against the light, respectively."

HOW SHOULD YOU MAKE THE MEASUREMENT OF THE LIGHT SPEED IN THESE TWO CASES? THEY NEVER GET ABOUT IT!

**20)**
***University Physics; Young, Freedmann;***

*p1243:*
"Einstein's conceptual leap was to recognize that if Maxwell's equations are valid in all inertial frames, then the speed of light in vacuum should be the same in all frames and in all directions"

An observer at rest against the light source and one in motion must measure the same light speed, *c*.
ONE NEVER SAY HOW TO MEASURE THE LIGHT SPEED!

**21)**
***The European Miracle; Book 1; 2016;
Robin T. Trnovsky***

*p274:*
"... the Croatian-Serbian genius Nikola Tesla's critique of relativity in, for example, the New York Times on July 11, 1935. Tesla claimed, among other things, that the theory of relativity is a magnificent mathematical costume that fascinates, dazzles and blind people to the underlying inaccuracies. He based his criticism in the first place on the fact that space cannot be curved for the simple reason that it cannot possess any properties, since it can only speak of characteristics when dealing with matter that fills space. Tesla said, 'To say that space is crooked in the presence of large bodies is like claiming that something can affect nothing.' There is undoubtedly a feeling that Tesla was strongly influenced by Parmenide's thoughts on 'nihil fit ex nihilo' from around 500 BC. "

## To physicists and mathematicians

To all physicists and mathematicians who have examined the special theory of relativity, I want to say the following:

I show in some parts of my research that:
- Lorentz transformations are incorrect, are not self-consistent.
- Michelson interferometer was **not suitable for use** to measure the Earth's velocity in space!
- parts of SR are presented with incorrect use of mathematics
- ... and more

This means that the special theory of relativity is also incorrect, is not self-consistent.

I might be wrong in any of these parts of my book. But it is enough if I am right in one of my claims to overthrow this whole theory.

The article *Special Relativity and Reality* (last in this book) is undoubtedly a proof that neither can be disputed nor ignored !!!

Quote from the book Kosmos - a short story
by Stephen Hawking:

*"A physical theory is always provisional in the sense that it is merely a hypothesis: one can never prove it. However many times the experimental results are consistent with a particular theory, one can never be sure that the results will not contradict the theory next time. On the other hand,* **one can disprove a theory by finding only one that does not correspond to the predictions of theory.** *"*

# Articles submitted for publication

# Article I sent to Physica Scripta

Physica Scripta is an international scientific journal for experimental and theoretical physics. It was established in 1970 and is published by IOP Publishing, endorsed by the Royal Swedish Academy of Sciences.

I have sent a number of articles to various magazines. Most I have sent to Physica Scripta:
No one was accepted

**19-Jun-2020**
**Lorentz Transformations And Time Dilation Do Not Verify Reality**

**10-May-2019**
**Special Relativity and Time Dilation!**

**25-Apr-2019**
**The Lorentz transformations and mathematics/11**

**17-Mar-2019**
**The Lorentz transformations and mathematics/10**

**15-Mar-2019**
**The Lorentz transformations and mathematics/09**

**02-Mar-2019**
**The Lorentz transformations and mathematics/08**

**17-Feb-2019**
**The Lorentz transformations and mathematics/07**

**16-Feb-2019**
**The Lorentz transformations and mathematics/06**

**10-Feb-2019**
**The Lorentz transformations and mathematics/05**

**11-Dec-2018**

**The Lorentz transformations and mathematics/04**

18-Nov-2018
**The Lorentz transformations and mathematics/02**

14-Nov-2018
**The special theory of relativity and reality**
10-Nov-2018
**The Lorentz transformations and mathematics/03**

03-Nov-2018
**The Lorentz transformations and mathematics/1**

02-Nov-2018
**Twin Paradox: ... and so they lived ... at the same age ... ever after**

18-Apr-2018
**The mathematics shows that the Lorentz transformations are not self-consistent**

16-Apr-2018
**Mathematics and Lorentz transformations**

## 08-May-2017
## Einstein's Theory of Special Relativity: A Mathematical Impossibility!

**No one was published!**

The answer I got from the magazine is basically the same for all articles:

*Thank you for your submission to Physica Scripta. We have assessed your manuscript and have considered its suitability for the journal very carefully. We regret to inform you that your article will not be considered for review as it does not meet our strict publication criteria.*

*The quality and presentation of any research published in Physica Scripta must be of the highest standard. Submissions should clearly demonstrate scientific rigour, extensive literature research and a careful assessment of the validity of any conclusions presented in the manuscript. Your manuscript does not meet these key publication criteria and we are unable to consider it further.*

Once I got the following answer:

*Please see the recent Editorial published in Issue 1 of 2016 (VOL 91) of the journal for more information of the new requirements for publishing in Physica Scripta:*
*http://iopscience.iop.org/article/10.1088/0031-8949/91/1/010401*

*Please note that only very significant specialised results are considered for publication in Physica Scripta, under the assumption that such results could have implications beyond their specialised field, and would be of interest to the readership of a broad scope journal. Therefore, specialised work with results only of interest to those working in a specific field, should be submitted to a topical journal.*

*If you would like to see a copy of the journal scope please visit the journal web page at http://iopscience.iop.org/1402-4896/page/Scope.*

*We are sorry that we cannot respond more positively and wish you luck in publishing your article elsewhere.*

# Article I sent to Physics Essays

**26 August 2019**
**Mathematics shows that**
**the Lorentz transformations are not**
**self-consistent**

Re: Manuscript: MATHEMATICS SHOWS THAT THE LORENTZ TRANSFORMATION ARE NOT SELF-CONSIDTENT, by Jan Slowak, submitted for publication in *Physics Essays* (received 26 August 2019).

Dear Mr. Slowak:
This is to acknowledge receipt of the above manuscript and to thank you for submitting it for publication in *Physics Essays*.
I will be pleased to examine the manuscript and inform as soon as possible about its suitability for the journal.
The timely review of your manuscript is a cooperative process between the Editor and you, the author. For an expedited processing of your manuscript, if you do not hear from me earlier, you are requested to inquire regularly in the future at reasonable time intervals (say, three months after last notification) about the status of your paper.

## 4 december 2019

After 3 and a half months, after 9 cover-letters, my article is accepted for publishing in *Physics Essays*.

Dear Mr. Slowak:

It is my pleasure to inform you that your paper is accepted for publication in *Physics Essays*. It will appear in the March 2020 issue (Vol. 33 No. 1) and you should receive the galley proofs for correction within a month.

I thank you for having submitted your work to us.

My best regards.

## Emilio Panarella

EP/ep

## 21 December 2019
## Lorentz Transformations - The Sound versus The Light

**First accepted for publication then denied publication!**

# Published articles

# 1)
# Physics Essays

## Mathematics shows that the Lorentz transformations are not self-consistent

http://physicsessays.org/browse-journal-2/product/1770-3-jan-slowak-mathematics-shows-that-the-lorentz-transformations-are-not-self-consistent.html

# 2)
# SCIREA Journal of Physics

## Lorentz Transformations And Time Dilation Do Not Verify Reality

http://www.scirea.org/journal/PaperInformation?PaperID=3699

## 3)
# SCIREA Journal of Physics

**Lorentz Transformations - The Sound versus The Light**

http://www.scirea.org/journal/PaperInformation?PaperID=3718

*Jan Slowak: Special Relativity is Nonsense*

# The Lorentz transformations and mathematics/01

## Jan Slowak

# 1 Abstract

Einstein's theory of special relativity is a generally accepted theory that analyses relationships between two inertial reference frames moving at a constant speed against each other. In this work, we analyze the derivation of Lorentz transformations only from the point of view of mathematics. This analysis shows that the Lorentz transformations are nonsense.

# 2 Keywords

Special Relativity, Lorentz transformations, Mathematics

# 3 Introduction

When studying a physical phenomenon, a mathematical model is developed to describ it. Such a model comprises built-in physical laws held together by mathematical tools. If the description of the physical phenomenon is correct, the mathematical model is also correct.

We consider Lorentz transformations, below.

$$x' = (x - vt)\gamma, \qquad (1)$$

$$t' = (t - \frac{vx}{c^2})\gamma \qquad (2)$$

where $\gamma = \frac{1}{\sqrt{1-\frac{v^2}{c^2}}}$ is called the Lorentz factor.

The Lorentz factor is a function of the velocity $v$.
If $v = 0$ then $\gamma = 1$ otherwise $\gamma > 1$. $\gamma$ is always $\neq 0$.
In *[1], pages 14-15*, they derive Lorentz transformations above and as a condition they have $v > 0$.

One derives Lorentz transformations (1), (2) from two linear general transformations / equations.

$$x' = Ax + Bt \tag{3}$$

$$t' = Cx + Dt \tag{4}$$

For this derivation one use three special cases:

$$x' = 0, x = vt \tag{5}$$

$$x' = -vt', x = 0 \tag{6}$$

$$x' = ct', x = ct \tag{7}$$

# 4 Mathematical demonstration

We briefly show how the derivation went to:
From (3), (5) $\Rightarrow 0 = Avt + Bt \Rightarrow Av + B = 0 \Rightarrow B = -Av$
We replace $B = -Av$ to (3) $\Rightarrow$

$$x' = A(x - vt) \tag{8}$$

We attach document [2] with three pictures.
Fig. 01 shows the situation when the thought experiment begins.
Both reference systems are then in the same point.

In Fig. 02, we show special case (5).
I would like to draw attention to the following:
The two linear equations (3), (4) should be applicable to all points in spacetime.
The two linear equations (1), (2) should be applicable to all points in spacetime.
Even the equation (8), which is an intermediate result, should be applicable to all points in spacetime.
We now look at Fig. 03 where we have an event that occurs at a distance $d$ from the S'-origo.
Then we have the following relationships:
$x' = d$ and $x = vt + d \Rightarrow x = vt + x' \Rightarrow$
$$x' = x - vt \qquad (9)$$
And now with simple mathematics we can say that from (8) and (9) $\Rightarrow d = Ad \Rightarrow A = 1$.
But the Lorentz transformations say that this factor $A$ is equal to the Lorentz factor, $\gamma$.
But we only have one reality and two results.
$\Rightarrow \gamma = 1 \Rightarrow v = 0$.

# 5 Conclusions

**The Lorentz transformations only applies when the two reference systems are stationary relative to each other! They are nonsense.**

From this follows that the theory of special relativity is nonsense.

# 6 References

[1] Modern Physics, second edition, Randy Harris, 2008, pages 14–15.
[2] Figures: Fig. 01, Fig. 02, Fig. 03

# Figures:
# LT and mathematics-1

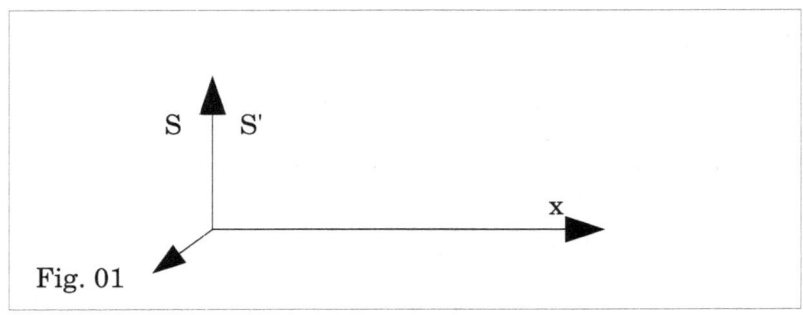

Fig. 01

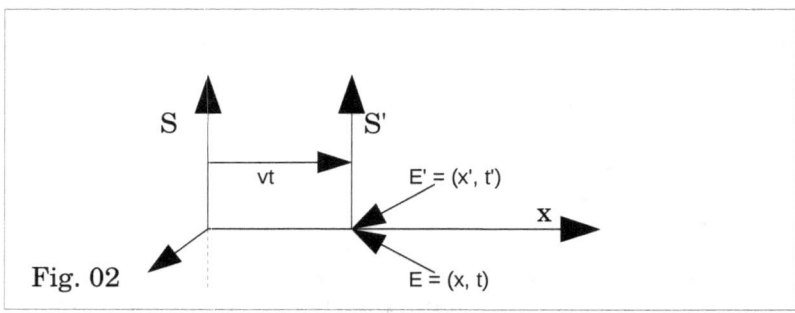

Fig. 02

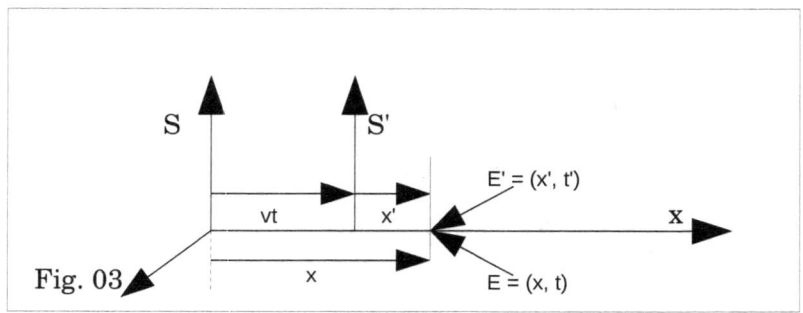

Fig. 03

# The Lorentz transformations and mathematics/02

## Jan Slowak

## 1 Abstract

Einstein's theory of special relativity is a generally accepted theory that analyses relationships between two inertial reference frames moving at a constant speed against each other. In this work, we analyze how time is registred in these two reference frames. This analysis shows that the time which is registered in reference frame in motion is contrary to the one claimed by the Lorentz transformations.

## 2 Keywords

Special Relativity, Lorentz transformations, Mathematics

## 3 Introduction

When studying a physical phenomenon, a mathematical model is developed to describ it. Such a model comprises built-in physical laws held together by mathematical tools. If the description of the physical phenomenon is correct, the mathematical model is also correct.

In this thought experiment, we consider two inertial reference systems, S and S', whose x-axes coincide with each other. At time $t = 0, t' = 0$, on the x-axis an event occurs, a light signal is transmitted to S, S'. We will analyze the time that the light signal needs to reach these two reference systems.

In Fig. 00 we see the basic elements of this thought experiment.

# 4 Partial experiment 1

See Fig. 01.

At time $t = 0, t' = 0$, on the x-axis an event occurs, a light signal is transmitted to S, S'. In this case, the two reference systems are in the same point and they are stationary relative to each other. The light signal fails the same distance, regardless of whether it is S or S'. The distance between S-origo, S'-origo and the point where the event occurs is the same.

$$x' = ct', x = ct \Rightarrow ct' = ct \Rightarrow t' = t$$

# 5 Partial experiment 2

See Fig. 02.

Now we look at the situation when S' moves relative S with constant velocity $v > 0$. Meanwhile the light signal moves towards S' and reaches this reference systems, moves S' to the right with a distance of $vt'$.

$$x = x' + vt' \Rightarrow ct = ct' + vt' \Rightarrow ct = (c+v)t' \Rightarrow t' = t\frac{c}{c+v}$$

# 6 Partial experiment 3

See Fig. 03.

What happens if the speed with which S' moves is bigger and bigger? What happens if this speed approaches the speed of light? We know that no material object, no reference systems can move with the speed of light. Therefore, in this thought experiment, we replace S' with a light signal which starts in S-origo, S'-origo.

Then the two light signals meet exactly in the middle of the distance between the S-origo and the point where the event occurred.

$x = ct' + ct' \Rightarrow ct = ct' + ct' \Rightarrow ct = 2ct' \Rightarrow t = 2t' \Rightarrow t' = \frac{t}{2}$

# 7 Summary

We summarize these three partial experiments. We look at the time recorded in S'. This time begins with having a value equal to $t$ and becomes less the higher the speed $v$ becomes.

$t' = t$ than decreases $t'$ until $t' = \frac{t}{2}$

This is contrary to the relationship between $t'$ and $t$, which originates from Lorentz transformations. From Lorentz transformations $\Rightarrow t' = t\gamma$

where $\gamma = \frac{1}{\sqrt{1-\frac{v^2}{c^2}}}$ is called the Lorentz factor.
If $v = 0$ then $\gamma = 1$ otherwise $\gamma > 1$. $\gamma$ is always $\neq 0$.

$\Rightarrow t' = t\gamma \Rightarrow t' > t$

From the analysis in this work we have that $\Rightarrow t' \leq t$!

How is this possible? I think the analyzes we did in this letter are quite logical. Where is the fault? We can not have such different results for one and the same phenomenon. No matter what thought experiments we do, no matter what mathematical, physical or logical tools we use, we should come to the same conclusion.

Where is the error in my analysis? If there is no error in my analysis then the error must be in the derivation of Lorentz transformations.

# 8 Conclusions

**The analysis of some thought experiments leads to the conclusion that the time recorded by reference system in motion is in contradiction with time indicated by Lorentz transformations. Is this analysis wrong or are Lorentz transformation not self-consistent?**

# 9 References

[1] Modern Physics, second edition, Randy Harris, 2008, pages 14–15.
[2] Figures: Fig. 00, Fig. 01, Fig. 02, Fig. 03

# Figures:
# LT and mathematics-2

Fig. 00

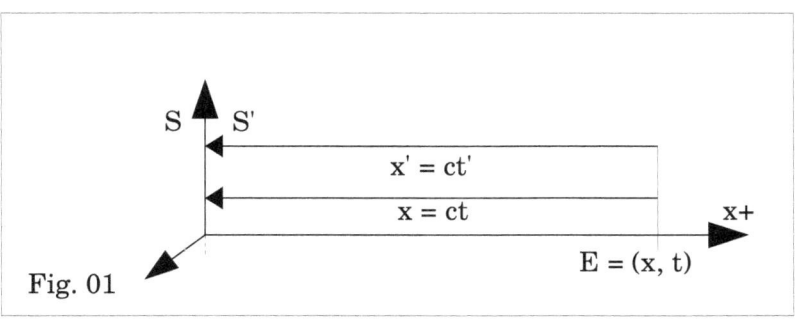

Fig. 01

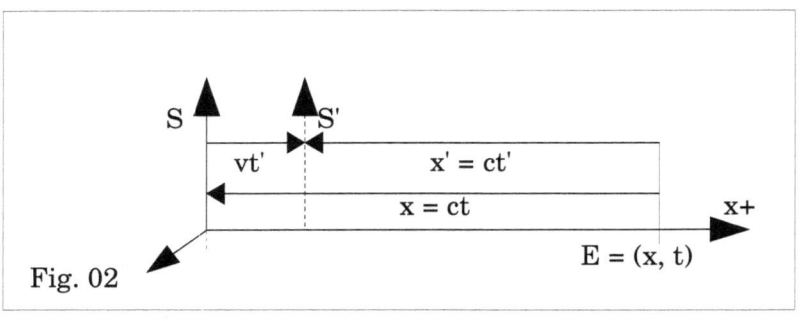

Fig. 02

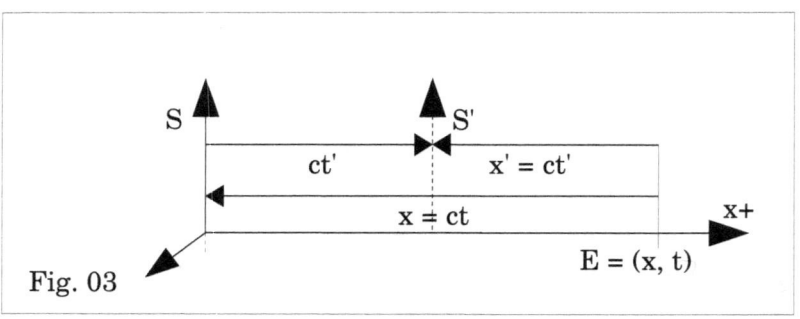

Fig. 03

# The Lorentz transformations and mathematics/03

## Jan Slowak

## 1 Abstract

Einstein's theory of special relativity is a generally accepted theory that analyses relationships between two inertial reference frames moving at a constant speed against each other. In this work, we analyze the derivation of Lorentz transformations only from the point of view of mathematics. We look at an equation not used in the derivation. This analysis shows that the Lorentz transformations are nonsense.

## 2 Keywords

Special Relativity, Lorentz transformations, Mathematics

## 3 Introduction

When studying a physical phenomenon, a mathematical model is developed to describ it. Such a model comprises built-in physical laws held together by mathematical tools. If the description of the physical phenomenon is correct, the mathematical model is also correct.

We consider Lorentz transformations, below.

$$x' = (x - vt)\gamma, \tag{1}$$

$$t' = (t - \frac{vx}{c^2})\gamma \qquad (2)$$

where $\gamma = \frac{1}{\sqrt{1-\frac{v^2}{c^2}}}$ is called the Lorentz factor.

The Lorentz factor is a function of the velocity $v$.
If $v = 0$ then $\gamma = 1$ otherwise $\gamma > 1$. $\gamma$ is always $\neq 0$.
In [1], pages 14-15, they derive Lorentz transformations above and as a condition they have $v > 0$.

One derives Lorentz transformations (1), (2) from two linear general transformations / equations.

$$x' = Ax + Bt \qquad (3)$$

$$t' = Cx + Dt \qquad (4)$$

For this derivation one use three special cases:

$$x' = 0, x = vt \qquad (5)$$

$$x' = -vt', x = 0 \qquad (6)$$

$$x' = ct', x = ct \qquad (7)$$

When one derive Lorentz transformations in [1], one get the following intermediate results:

$$B = -Av, C = -Av/c^2, D = A, A = \gamma \qquad (8)$$

As one of the key conclusion of SR is the time dilation in the reference frame that is in motion.

$$t' = t\gamma \qquad (9)$$

# 4 Mathematical demonstration

I would like to draw attention to the following:
During the derivation in [1] they do not use combination (4), (5). Why?
We look at this now: From (4), (5) $\Rightarrow t' = Cvt + Dt$
We are using now (8)
$\Rightarrow t' = (-Av/c^2)vt + At \Rightarrow t' = At(1 - v^2/c^2) \Rightarrow t' = At/\gamma^2$
$\Rightarrow t' = \gamma t/\gamma^2 \Rightarrow t' = t/\gamma$
This results in

$$t' = t/\gamma \qquad (10)$$

If we now equate (9) and (10) $\Rightarrow t\gamma = t/\gamma \Rightarrow t(\gamma - \frac{1}{\gamma}) = 0 \Rightarrow t = 0$ or $v = 0$.

This means that (2) and special case (5) only applies to $t = 0$ or $v = 0$.

When $t = 0$, the two reference systems are in one and the same point.
The condition $v = 0$ is in contradiction with the original condition of experiment which states that the two reference systems move with the velocity $v > 0$ relative to each other.

# 5 Conclusions

**The Lorentz transformations only applies when the two reference systems are in the same point or thay are not self-consistent! They are nonsense.**

From this follows that the theory of special relativity is nonsense.

# 6  References

[1] Modern Physics, second edition, Randy Harris, 2008, pages 14–15.

# The Lorentz transformations and mathematics/04

## Jan Slowak

## 1 Abstract

Einstein's theory of special relativity is a generally accepted theory that analyses relationships between two inertial reference frames moving at a constant speed against each other. In this work, we analyze the derivation of Lorentz transformations only from the point of view of mathematics. This analysis shows that the Lorentz transformations are nonsense.

## 2 Keywords

Special Relativity, Lorentz transformations, Mathematics

## 3 Introduction

When studying a physical phenomenon, a mathematical model is developed to describ it. Such a model comprises built-in physical laws held together by mathematical tools. If the description of the physical phenomenon is correct, the mathematical model is also correct.

We consider Lorentz transformations, below.

$$x' = (x - vt)\gamma, \tag{1}$$

$$t' = (t - \frac{vx}{c^2})\gamma \tag{2}$$

where $\gamma = \frac{1}{\sqrt{1-\frac{v^2}{c^2}}}$ is called the Lorentz factor.

The Lorentz factor is a function of the velocity $v$.
If $v = 0$ then $\gamma = 1$ otherwise $\gamma > 1$. $\gamma$ is always $\neq 0$.
In [1], pages 14-15, they derive Lorentz transformations above and as a condition they have $v > 0$.

One derives Lorentz transformations (1), (2) from two linear general transformations / equations.

$$x' = Ax + Bt \tag{3}$$

$$t' = Cx + Dt \tag{4}$$

For this derivation one use three special cases:

$$x' = 0, x = vt \tag{5}$$

$$x' = -vt', x = 0 \tag{6}$$

$$x' = ct', x = ct \tag{7}$$

When one derive Lorentz transformations in [1], one get the following intermediate results:

$$B = -Av, C = -Av/c^2, D = A, A = \gamma \tag{8}$$

We look at how this derivation went to.

# 4 Mathematical demonstration

From (3) and (5) results $B = -Av \Rightarrow$
that from $x' = Ax + Bt$ and $x' = 0, x = vt$ results $B = -Av$.
This means that $B = -Av$ applies to $x' = 0$!!!

From (3), (4) and (6) results $B = -Dv \Rightarrow$
that from $x' = Ax + Bt, t' = Cx + Dt$ and $x' = -vt', x = 0$ results $B = -Dv$.
This means that $B = -Dv$ applies to $x = 0$!!!

Furthermore, if $B = -Av$ and $B = -Dv$ then $\Rightarrow D = A$. And that's right. But this means that the result $D = A$ applies to $\{x' = 0, x = 0\}$.

This means that the result $D = A$ applies only to the moment when the two reference systems are in the same point.

# 5 Conclusions

**The Lorentz transformations only applies when the two reference systems are in the same point! They are nonsense.**

This means that the theory of special relativity is nonsense.

# 6 References

[1] Modern Physics, second edition, Randy Harris, 2008, pages 14–15.

# Special Relativity and Reality

## Jan Slowak

## 1 Abstract

Einstein's theory of special relativity is a generally accepted theory that analyses relationships between two inertial reference frames moving at a constant speed against each other. In this work we compare two thought experiments:
1) when two reference systems are stationary relative to each other
2) when two reference systems are moving relative to each other

## 2 Keywords

Special Relativity, Lorentz transformations, Reality

## 3 Introduction

**Quote:**
"At the same time, these are the worlds we have the hardest to understand, worlds where illustrative models deceive us and we find paradoxes. But there is only one world and it has no paradoxes. Only our models that hold paradoxes "[1]

**My motto:**
When we study physical phenomena, we always make a mathematical model of them. In such a model there are built-in physical laws that are held together by mathematical tools. If the description of the physical phenomenon is correct, the mathematical model is also correct!

We consider the following thought experiments:

Two reference systems S and S' are in the same point. Their x-axes coincide with each other. On this axis there are two points:
- $x_d$ located at a distance $d$ from S-origo (S'-origo)
- $x_D$ located at a distance $D$ from S-origo (S'-origo)

See Fig. 01.

# 4 Partial experiment 1

We place S' in the point $x_d$. When this thought experiment begins, clocks are reset in both reference systems, the time begins to count from zero, $t = 0, t' = 0$. Then an event occurs at the point $x_D$, a light signal is sent to S, S'. We denote this event with $E = (x, t), E' = (x', t')$. Note that there is **only one event** but we denote it with two different notations, one for each reference system.

The two reference systems get knowledge of the event $E, E'$ when the light signal reaches the respective system. The light signal moves at the speed of light $c$.
See Fig. 02.

Then we can calculate the following:
1) $t' = \frac{D-d}{c}$, the time the light signal needs to pass the distance between the points $x_D$ and $x_d$
2) $t = \frac{D}{c}$, the time the light signal needs to pass the distance between the point $x_D$ and the S-origo
3) the distance between S'-origo and S-origo is $d$, the time the light signal needs to pass this distance is $\frac{d}{c}$.

We consider 3 (three) points on the x-axis:
1) S-origo ($x = 0$)
2) S'-origo (same as $x_d$)

3) the point where the event occurs (same as $x_D$)

We consider three distances created by three points above:
1) distance $[S - origo, S' - origo] = [0, x_d]$ has the length $d = x_d$
2) distance $[S' - origo, E(E')] = [x_d, x_D]$ has the length $D - d = x_D - x_d$
3) distance $[S - origo, E(E')] = [0, x_D]$ has the length $D = x_D$

**I hope so far there is nothing that a physicist/mathematician would not accept.**

Consider Fig. 02 again. Keep in mind that the three points are fixed. Keep in mind that **the light signal moves independently of the source and observer's own motion!** So it does not matter, it does not matter at all, if S' is there, if S' is stagnant relative to S or if S' is in motion relativ to S. **The light signal pass the three distances in the same way and during the same time.**

# 5 Partial experiment 2

Now we repeat the above experiments with the following modification.
1) When the event $E, E'$ occurs, both reference systems are in the same point, $x = 0$
2) S' will move to the right with speed $v = \frac{d}{t'} = \frac{d}{\frac{D-d}{c}}$

This means that S' will reach the point $x_d$ at the same moment when the light signal reaches this point.

See Fig. 03.

Consider these two pictures, Fig. 02 and Fig. 03, again.

At the moment the light signal reaches S', **at that moment**, there is no difference between these two pictures. There are no differences! We have the same distances between the three fixed points. The light signal needs same time to pass these distances.

We generalize this, see Fig. 04.

We have the following relations:

$x = vt' + x'$ or $x' = x - vt'$

We have only one reality and this tells us that the following relation applies:

$x = x' + vt'$

If we look at Lorentz's transformations, there are the following:

$x = (x' + vt')\gamma$, where $\gamma = \frac{1}{\sqrt{1-\frac{v^2}{c^2}}}$ is called the Lorentz factor.

If we equate these two relations between $(x, t)$ and $(x', t')$ we get

$\gamma = 1 \to v = 0$

# 6 Summary

We have two models that come into two different relationships between the same physical quantities. **But we have only one reality!**
One of these two relationships must be wrong!
I think that the derivation I made contains no errors in terms of physics, mathematics or logic.
The derivation of Lorentz transformations in [2] and [3] is "not self-consistent".

Lorentz transformations do not verify reality. Lorentz transformations are nonsense!

# 7 Conclusion

Because of this, the whole special theory of relativity is nonsense!

# 8 References

[1] At the crossroads: essays about man and her future; Ulf Sinnerstad; 2006; swedish
[2] The special and general theory of relativity; Albert Einstein; The first part; About the special theory of relativity; 2006; swedish
[3] Modern Physics; Second edition; Randy Harris; Chapter 2; Special Relativity; 2008

# Figures:
# Special Relativity and Reality

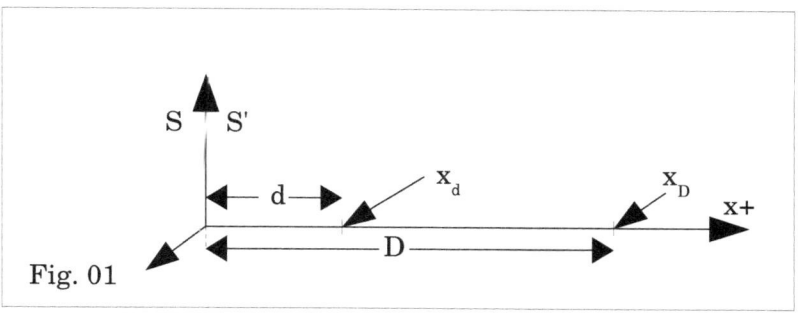

Fig. 01

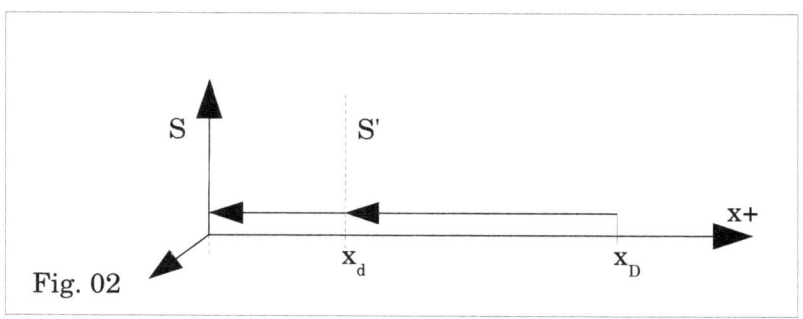

Fig. 02

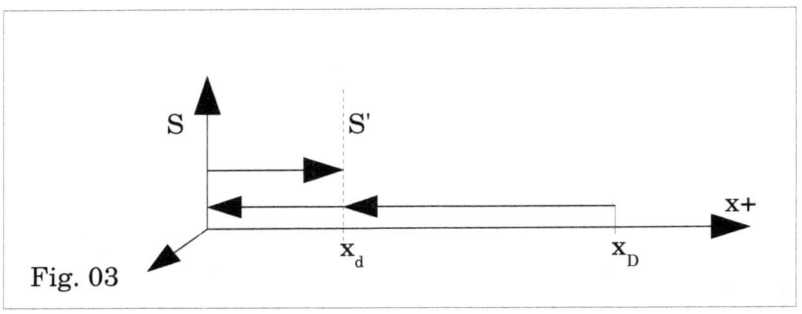
Fig. 03

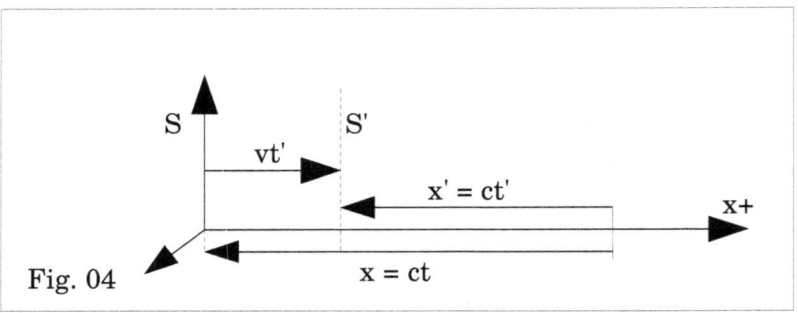
Fig. 04

www.ingramcontent.com/pod-product-compliance
Lightning Source LLC
Chambersburg PA
CBHW071206240526
**45470CB00018B/1510**